国外优秀数学著作
原版系列

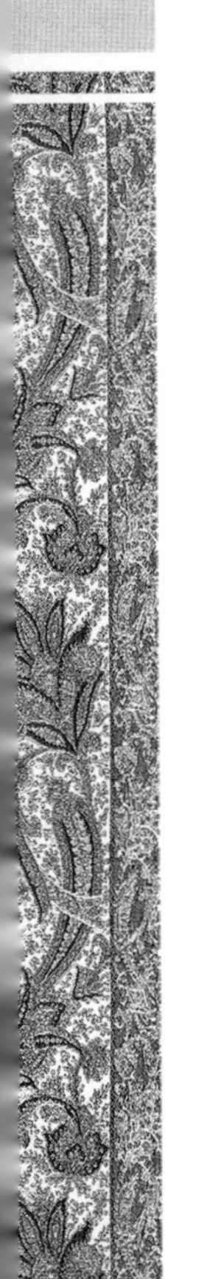

一种基于混沌的非线性最优化问题
——作业调度问题

A Chaos Based Approach for Nonlinear Optimization Problems
—On the Job Shop Scheduling Problem

（英文）

[埃] M.A. 艾尔—萨尔巴吉（M.A. El-Shorbagy）
[埃] S. 纳斯尔（S. Nasr）
[埃] 阿卜杜拉·A. 穆萨（Abd allah A. Mousa） 著

HITP
哈爾濱工業大學出版社
HARBIN INSTITUTE OF TECHNOLOGY PRESS

黑版贸审字 08－2019－180 号

图书在版编目(CIP)数据

一种基于混沌的非线性最优化问题:作业调度问题＝A Chaos Based Approach for Nonlinear Optimization Problems:On the Job Shop Scheduling Problem:英文/(埃)M. A. 艾尔－萨尔巴吉(M. A. El-Shorbagy),(埃)S. 纳斯尔(S. Nasr),(埃)阿卜杜拉·A. 穆萨(Abd allah A. Mousa)著. —哈尔滨:哈尔滨工业大学出版社,2023. 3

ISBN 978-7-5767-0678-9

Ⅰ. ①一… Ⅱ. ①M… ②S… ③阿… Ⅲ. ①非线性－最优化算法－英文 Ⅳ. ①O224

中国国家版本馆 CIP 数据核字(2023)第 032540 号

YIZHONG JIYU HUNDUN DE FEIXIANXING ZUIYOUHUA WENTI:ZUOYE DIAODU WENTI

策划编辑 刘培杰 杜莹雪
责任编辑 张永芹 邵长玲
封面设计 孙茵艾
出版发行 哈尔滨工业大学出版社
社　　址 哈尔滨市南岗区复华四道街 10 号 邮编 150006
传　　真 0451－86414749
网　　址 http://hitpress.hit.edu.cn
印　　刷 哈尔滨久利印刷有限公司
开　　本 886 mm×1 230 mm 1/32 印张 6.5 字数 137 千字
版　　次 2023 年 3 月第 1 版 2023 年 3 月第 1 次印刷
书　　号 ISBN 978-7-5767-0678-9
定　　价 38.00 元

Acknowledgements

First and foremost all gratitude and thanks to Allah almighty for his countless blessing and guidance, who inspired me to bring forth to light the material covered in this thesis.

I would like to express my sincere gratitude and appreciation to, **Dr. Abd Allah Abd Allah Mousa,** Assistant Professor of Engineering Mathematics, Department of Basic Engineering Sciences, Faculty of Engineering, Shebin El-Kom, Menoufiya University, for introducing the topic to me and sharing ideas. I am grateful for his guidance, advice and encouragement throughout the course of this work.

I would also like to express my sincere gratitude and appreciation to, **Dr. Islam Mohamed Ibrahim El- Desoky,** Assistant Professor of Engineering Mathematics, Department of Basic Engineering Sciences, Faculty of Engineering, Shebin El-Kom, Menoufiya University, for his great supervision, support and encouragement during this work. His guidance helped me in all the time of research.

My gratitude is extended to, **Dr. Zeinab Mohamed Hendawy,** Lecturer of Engineering Mathematics, Department of Basic Engineering Sciences, Faculty of Engineering, Shebin El-Kom, Menoufiya University, for her help, guidance and advice. I am grateful to her for her time which she has given me and her faith and trust which she had given me during my work.

I am also hugely appreciative to **Dr. M. A. El-Shorbagy,** Lecturer of Engineering Mathematics, Department of Basic Engineering Sciences, Faculty of Engineering, Shebin El-Kom, Menoufiya University, for his guidance during my research. His support and inspiring suggestions have been precious for development of this thesis content.

I also acknowledge the support and facilities provided by my department, Department of Basic Engineering Sciences, Faculty of Engineering, Shebin El-Kom, Menoufiya University. I am also grateful to my family for giving birth to me at the first place and supporting me spiritually throughout my life their patience and faith in me.

S. Nasr

List of Abbreviations

ABC	Artificial bee colony
ACO	Ant Colony Optimization
ALPSO	Augmented Lagrange Particle Swarm Optimization
BKS	Best Known Solution
CEC'2005	Congress on Evolutionary Computation
CGA	Chaotic Genetic Algorithm
CHC	Cross-generational elitist selection, Heterogeneous recombination and Cataclysmic mutation
CLS	Chaotic local search
CX	Cycle Crossover
DE-Bin	Differential Evolution- binomial
DE-Exp	Differential Evolution- Exponential
EA	Evolutionary Algorithm
FT	Fisher and Thompson
GA	Genetic Algorithm
HGA	Hybrid Genetic Algorithm
HS	Harmony search
Ipop-GMA-ES	Increase Population-Covariant Matrix Evolutionary Strategy
JSSP	Job Shop Scheduling Problem
LA	Lawrence
LPP	linear Programming Problem
NLPP	Nonlinear Programming Problem
NP	Non-Deterministic Polynomial
OR	Operations Research
OX	Order Crossover
PMX	Partial-Mapped Crossover
PSO	Particle Swarm Optimization
SA	Simulated Annealing
SPT	shortest processing time rule
SaDE	Self-adaptive Differential Evolution
SQP	Sequential Quadratic Programming
SSGA	Steady –State Genetic Algorithm

List of Abbreviation

SQP	Sequential Quadratic Programming
SSGA	Steady –State Genetic Algorithm
SS-Arit	Scatter Search-Arithmetical
SS-BLX	Scatter Search- Blend Crossover
TS	Tabu Search

Contents

Page

Contents

page

Contents

Page

List of Figures

List of Figures

List of Tables

List of Tables

Abstract

Optimization problems are essentially research topic in many science and engineering disciplines and there are still many open questions in this area. Production scheduling is an essential factor of the logistical performance of production organizations. The importance of a production scheduling optimization is growing due a fact that most of the companies developed its production management style based on more suitable solutions. Job shop scheduling problem (JSSP) is a branch of production scheduling, which is among the hardest combinatorial optimization problem.

Variety techniques are appeared to solve optimization problems. Genetic algorithm is one of optimization techniques that enjoy an increasing interest in the optimization community. Chaos is a kind of universal nonlinear phenomena in all areas of science. In the recent years, chaos theory has been applied to many aspects of the optimization sciences.

In this thesis, we present a new hybrid optimization algorithm for solving one of the most important optimization problems) nonlinear optimization problems) . The hybrid algorithm is a combination between genetic algorithm and chaos theory. The integration between genetic algorithm and chaos local search procedures should offer the advantages of both optimization methods and increase the convergence to reach to the global solution. Our approach is tested on a set of standard instances taken from the literature.

In addition, hybrid genetic algorithm is presented to solve one of real optimization problems (job shop scheduling problem). Our algorithm is a combination between genetic algorithm and local search. We design a generation alternation model using genetic algorithm. Then, we applied local search based on the neighborhood structure in the genetic algorithm result. The proposed approach for solving job shop scheduling problem is tested on a set of standard instances taken from the literature. The computation results have validated the effectiveness of our proposed algorithm.

This thesis consists of six main chapters. These chapters can be described in the following manner:

CHAPTER 1: In this chapter a survey on related topics of optimization First, we propose mathematical model of optimization problems. Then, a classification of optimization Problems is introduced. Also, we propose in this chapter optimization techniques for solving optimization problems.

Abstract

CHAPTER 2: This chapter aims to introduce the working principles of genetic algorithm and explain how the genetic algorithm is applied for solving optimization problems. Also, we introduce in this chapter the genetic algorithm parameters and the advantages and disadvantages of using genetic algorithm for optimization problems.

CHAPTER 3: In this chapter, A new algorithm is proposed to solve nonlinear optimization problems. A new algorithm is a combination between one of optimization techniques (genetic algorithm) and chaos theory to enhance the performance and reaching to the optimal solution. It is a new algorithm that combines genetic searching features and chaos searching features for solving nonlinear optimization problems. Various kinds of benchmark problems have been tested to illustrate the successful result in finding optimal solution.

CHAPTER 4: In this chapter, we propose structure of job shop scheduling problem. Then, the job shop scheduling problem formulation is introduced. Also we show the complexity of job shop scheduling problem. Finally in this chapter, we introduce the techniques for solving job shop scheduling.

CHAPTER 5: This chapter intends to implement our new approach for solving job shop scheduling problem and explained it in detail. The experimental results for various kinds of benchmark problems that have been tested are discussed. The results are compared with another approach to show the reliability of our approach and its ability for solving job shop scheduling problem.

CHAPTER 6: This chapter describes some concluding remarks, recommendations and some points for further researches.

CHAPTER 1

A Survey on Related Topics

1.1 Introduction

Optimization is an art of selecting of the best alternative amongst a given set of options. Our lives are full of many optimization examples, what time do we get up in the morning so that we maximize the amount of sleep yet still make it to work on time? What is the best route to work? Which project do we tackle first? In the real world, there are many problems which should be optimized. Thus optimization is considered as a very important research topic for both scientists and engineers and still there are many open questions in its area [1-3].

Optimization is defined as maximization or minimization of a real function by systematical choose of input values from within an allowed set. Optimization problems are classified to large types and there is no single method available for solving all optimization problems efficiently [2].

The existence of optimization methods can be traced to the days of Newton, Lagrange, and Cauchy in the seventeenth century. Their first techniques could not solve different types of optimization problems and have limited scope in practical applications. By the middle of the twentieth century, the high-speed digital computers help to appear some numerical methods. This advancement also resulted in the emergence of several well defined new areas in optimization theory. Some of these new areas are linear programming, integer programming, quadratic programming, nonlinear programming, stochastic programming, separable programming, multiobjective programming, …, etc [4, 5]. Although these techniques tried to solve optimization problems in practical applications, there are many difficulties and limitations when they are applied.

Recently, some optimization methods that are conceptually different have been appeared labeled as advanced (modern) optimization techniques. The advanced optimization techniques are emerging as popular methods for the solution of complex engineering problems [6, 7]. Advanced techniques overcome difficulties and limitations of classical techniques.

The most important aim of this chapter is to introduce optimization problem model, classification of optimization problems and review of techniques used to solve optimization problems.

1.2 Mathematical Model of Optimization Problems

The main components of an optimization problem [3] are:

a) Objective Function

An objective function expresses one or more quantities which are to be minimized or maximized. The optimization problems may have a single objective function or more objective functions.

b) Variables

A set of unknowns, which are essential are called variables. The variables are used to define the objective function and constraints. One cannot choose design variable arbitrarily, they have to satisfy certain specified functional and other requirements.

c) Constraints

A set of constraints are those which allow the unknowns to take on certain values but exclude others. They are conditions that must be satisfied to render the design to be feasible.

The optimization problem must be solved for the values of the variables that satisfy constrains and minimize or (maximize) the objective function.

- **Statement of an optimization problem**

The general statement of optimization problem is:

Find x which minimizes or (maximizes) $f(x)$

$$\text{subject to:} \quad g_j \le 0 \qquad \text{for} \quad j = 1,...,m \tag{1.1}$$

$$h_i(x) = 0 \qquad \text{for} \quad i = 1,...,t; \tag{1.2}$$

where $f, g_1,..., g_j, h_1,..., h_i$ are functions defined on R^n, $x \in R^n$ and is a vector of n components $x_1,..., x_n$ [3]. The function f is the objective function, or the criterion function. If there are no constraints, problem is called unconstrained problems. If there are constraints, problem is called constrained problem and each of the constraints $g_j \le 0$ is called an inequality constraint, and each of the constraints $h_i(x) = 0$ is called equality constrain. At solving optimization problem, we are looking for a global solution and not stock on a local solution. Definition 1.1 introduces the difference between a local solution and global solution. Figure 1.1 illustrates this definition.

Definition 1.1

Let $x \in R^n$ and $x = (x_1, x_2,..., x_n)$ be a feasible solution to a minimization problem with objective function $f(x)$ [3]. We call x:

- A global minimum if $f(x) \le f(y)$ for every feasible point $y = (y_1, y_2,..., y_n)$.
- A local minimum if $f(x) \le f(y)$ for every feasible point $y = (y_1, y_2,..., y_n)$ sufficiently close to x.

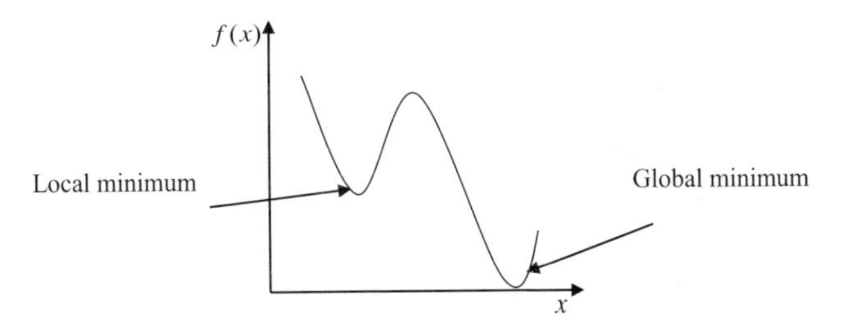

Figure 1.1 Global minimum and Local minimum

1.3 Classification of optimization problems

Optimization problems are classified according to several categories. The classification can be based on existence of constraints, nature of design variables, physical structure of the problem, nature of the equations involved, permissible value of the design variables, deterministic nature of the variables, separability of the functions and number of objective functions. None of these classification or their branches are necessarily mutually exclusive [3].

1.3.1 Classification based on existence of constraint

There are two broad categories of classification within this classification [3]:

a) Constrained optimization problem

One or more constraints exist in optimization problem.

b) Unconstrained optimization problems

No constraints exist in the optimization problems.

1.3.2 Classification based on nature of the design variables

There are two categories within this classification:

a) Static optimization problems

Static optimization problem is the problem in which the objective is finding a set of design parameters that make the objective function of these parameters minimum or maximum subject to certain constraints [3].

b) Dynamic optimization problems

Dynamic optimization problem is the problem in which the optimization strategy is based on partition the problem into smaller sub-problems. The objective is finding a set of design parameters, which are all continuous functions of some other parameters, that minimizes a prescribed function subject to a set of constraints [3].

1.3.3 Classification based on physical structure of the problem

Optimization problems can be classified as optimal control and non-optimal control optimization problems.

a) Optimal control optimization problem

In an optimal control problem, there are two types of variables control or design variables and state variables. The problem is to find the set of control or design variables such that the total objective function is minimized subject to constraints on the control and state variables.

$$\text{Find } x \text{ which minimizes } f(x)=\sum_{i=1}^{l} f_i(x_i,y_i) \tag{1.3}$$

$$\text{subject to: } q_i(x_i,y_i)+y_i=y_{i+1} \qquad i=1,2,\ldots,l \tag{1.4}$$

$$g_j(x_j)\le 0 \qquad j=1,2,\ldots,l \tag{1.5}$$

$$h_k(y_k)\le 0 \qquad k=1,2,\ldots,l; \tag{1.6}$$

where x_i is the control variable, y_i is the control variable, and f_i is the contribution of the stage to the total objective function; g_j, h_k and q_i are functions of x_i, y_i and l is the total number of stages [3].

b) Non-optimal control optimization problems

In non-optimal control problem, the variables are not control or design variables.

1.3.4 Classification based on nature of the equations involved

According to the nature of expressions for the objective function and the constraints, optimization problems are classified as linear, nonlinear, geometric and quadratic programming problems.

a) Nonlinear programming problem

The nonlinear programming problem is called a nonlinear programming (NLP) problem if any of functions among the objectives and the constraint functions is nonlinear. NLP problem is the most general form of optimization problem [3].

b) Geometric programming problem

In geometric programming problem, the objective function and constraints are expressed as polynomials in x [3].

c) Quadratic programming problem

Quadratic programming problem is a NLP problem with a quadratic objective function and linear equalities and inequalities constraints [3]. It is formulated as follows:

$$\text{Minimizes } f(x) = c + \sum_{i=1}^{n} q_i x_i + \sum_{i=1}^{n} \sum_{j=1}^{n} Q_{ij} x_i x_j \tag{1.7}$$

$$\text{subject to: } \sum_{i=1}^{n} a_{ij} x_i = b_j, \quad j = 1, 2, \cdots, m \tag{1.8}$$

$$x_i \geq 0, \qquad i = 1, 2, \cdots, n; \tag{1.9}$$

where c, q_i, Q_{ij}, a_{ij} and b_j are constants.

d) Linear programming problem

The optimization problem is called a linear programming problem (LPP) if the objective function and all the constraints are 'linear' functions. LPP is often stated in the standard form [3]:

$$\text{Find } x^* = \begin{Bmatrix} x_1^* \\ x_2^* \\ \vdots \\ x_n^* \end{Bmatrix}; \text{ which}$$

$$\text{minimizes } f(x) = \sum_{i=1}^{n} c_i x_i \tag{1.10}$$

$$\text{subject to: } \sum_{i=1}^{n} a_{ij} x_i = b_j, \quad j = 1, 2, \cdots, m \tag{1.11}$$

$$x_i \geq 0, \qquad i = 1, 2, \cdots, n \tag{1.12}$$

where c_i, a_{ij} and b_j are constants.

1.3.5 Classification based on permissible values of the design variables

Under this classification, problems are classified as integer and real-valued programming problems.

a) Integer programming problems

The problem is called an integer programming problem if some or all of the design variables of an optimization problem are confined to take only integer values. As example, the optimization problem is to find number of articles needed for an operation with least effort. Thus, the objective function is minimization of the effort required for the operation and the decision variables, i.e. the number of articles used, can take only integer values [3].

b) Real valued programming problems

A real-valued problem is the problem in which the objective function is to minimize or maximize a real function by systematically choosing the values of real variables from within an allowed set. If the allowed set contains only real values the problem is called a real-valued programming problem [3].

1.3.6 Classification based on deterministic nature of the variables

Under this classification, problems are classified as stochastic programming problem and deterministic programming problem.

a) Stochastic programming problem

In stochastic programming problem, some or all the design variables are non-deterministically or stochastically expressed. As example estimation of life span of structures which have probabilistic inputs of the concrete strength and load capacity. This problem is a stochastic programming problem as one can only estimate stochastically the life span of the structure [8].

b) Deterministic programming problem

In deterministic optimization, it is assumed that the data for the given problem are known accurately. However, for many actual problems, the data cannot be known accurately for a variety of reasons. The first reason is due to simple measurement error. The second and more fundamental reason is that some data represent information about the future (e. g., product demand or price for a future time period) and simply cannot be known with certainty.

1.3.7 Classification based on separability of the functions

According to the separability of the objective and constraint functions, optimization problems are classified as separable and non-separable programming problems.

a) Separable programming problems

If function $f(x)$ can be expressed as the sum of n single variable functions $f_1(x_1), f_2(x_2), ..., f_n(x_n)$ the problem is said to be separable programming problem. In separable programming problem, the objective function and the constraints are separable [20]. Separable programming problem can be expressed as:

$$\text{Find } x \text{ which minimizes } f(x) = \sum_{i=1}^{n} f_i x_i \tag{1.13}$$

$$\text{subject to: } g_j(x) = \sum_{i=1}^{n} g_{ij}(x_i) \le b_j, \quad j = 1, 2, \cdots, m; \tag{1.14}$$

where b_j is constant.

b) Non-separable programming problems

In non-separable programming problem, the objective function and the constraints are non-separable.

1.3.8 Classification based on number of the objective functions

According to this classification, optimization problems are classified as single and multiobjective programming problems.

a) Single objective programming problem

In single-objective programming problem, there is only a single objective.

b) Multiobjective programming problem

In multiobjective programming problem, there is more than one objective. Usually the different objectives are not compatible. The variables that optimize one objective may be far the solution from optimal for the others. With multi-objectives, the problem can be reformulated as single objective problems by either forming a weighted combination of the different objectives or by treating some of the objectives as constraints [9]. A multiobjective programming problem is stated as follows:

Find x which minimizes $f_1(x), f_2(x), ..., f_n(x)$

$$\text{subject to: } g_j(x) \le 0, \qquad j = 1,2,...,m; \tag{1.15}$$

where $x \in R^n$ and the objective functions are required to be minimized simultaneously.

1.4 Optimization Techniques

Optimization problems are classified to large types. There is no universal optimization routine which will solve any given problem more efficiently than any other. Various classical methods have appeared to solve optimization problems [11-13]. Classical -techniques have some difficulties and limitations. The main problem is to solve non-differentiable functions. Moreover, classical techniques often fail to solve optimization problems that have many local optima. Classical techniques generally fail to solve such large-scale problems especially with nonlinear objective functions. In recent years, major advances in optimization occurred. Other new techniques appeared to solve any optimization problem called advanced techniques. These techniques are used to solve linear, nonlinear, differential and non-differential optimization problems. The advanced techniques are suitable for practical applications that are often nonlinear or non-differential [11].

1.4.1 Classical Optimization Techniques

The first classical optimization techniques works by deriving the equations with respect to each parameter in turn, setting the set of partial differential equations to zero, and solving this set of simultaneous equations. In practical applications, these first techniques are not suitable. Practical applications may be not continuous equations, impossible to differentiate the equations, non-linear equations may be obtained. A large variety of other classical optimization techniques appears tried to deal with practical applications [4, 11]. These techniques are as linear programming, integer programming, quadratic programming, nonlinear programming, stochastic programming, separable programming problem, multiobjective programming,…,etc. Here, we concern on nonlinear programming [12, 13].

1.4.1.1 Nonlinear Programming

In (NLP) problem, any of the functions among the objectives and constraint functions is nonlinear. The NLP techniques are treated as the most general techniques for the solution of any optimization problem. Finding the minimum of a nonlinear function is especially difficult. For NLP, the search algorithm can be classified into two broad categories as direct search methods and gradient methods. The direct search methods require only the objective function values but not the partial derivatives of the function in finding the minimum and hence are often called the nongradient methods. These methods are most suitable for simple problems involving a relatively small number of variables. The gradient techniques require, in addition to the function values, the first and in some cases the second derivatives of the objective [3, 12].

A) Newton's Method

Newton's method is an indirect technique for solving unconstrained optimization problems that generates a sequence of iterates x_k. The next iteration x_{k+1} is obtained by adding a step d_k. The step d_k is computed by solving the following sub problem:

$$\text{Minimize } m_k(d)=f(x_k)+\nabla f(x_k)^T d+\frac{1}{2}d^T\nabla^2 f(x_k). \tag{1.16}$$

Finally, we set $x_{k+1}=x_k+d_k$.

Newton's method converges quadratically when started close enough to a Nonsingular local minimizer (positive definite Hessian). The disadvantage of Newton's method is not globally convergent to stationary point [13].

To solve the equality constrained optimization problem, there are different methods developed for this purpose. The following subsections show some of these methods,

For example, the Penalty method, the augmented Lagrangian method, and the sequential quadratic programming (SQP) method.

B) Penalty Method

Penalty methods are an indirect method and are a certain class of algorithms to solve constraint optimization problems. The Penalty method replaces a constraint optimization problem by a sequence of unconstrained optimization problems whose solutions must convergence to the solution of the original constrained problem.

The Penalty functions that appear more frequently in the literature for equality constrained optimization problem are the quadratic Penalty function

$$Q(x;r)=f(x)+\frac{r}{2}\|h(x)\|^2, \tag{1.17}$$

where $r>0$ is the Penalty parameter. Usually, Penalty methods generate a sequence of infeasible points and under very mild conditions any limit point is an optimal solution to the original problem. In most of the methods, the feasibility is achieved only at the solution. However, there are Penalty functions that generate a sequence of points that may be interior or exterior to the feasible set [14-18].

It is well known that, in order to guarantee convergence of the penalty methods, the penalty parameter must go to infinity, and so, the problem becomes increasingly ill conditioned. This is the motivation for introducing the method of multiplier or the augmented Lagrangian method.

C) Augmented Lagrangian Method

The augmented Lagrangian method is an indirect technique and also known as the method of multiplier. The motivation for the augmented Lagrangian methods came from the desire of avoiding ill conditioning associated with usual penalty methods. Indeed, in contrast to penalty methods, the penalty parameter need not to go to infinity to achieve convergence of the augmented Lagrangian methods. As a consequence, the augmented Lagrangian has a good conditioning, and the methods are robust for solving nonlinear programs.

Let us consider the augmented Lagrangian function $\psi:\Re^n\times\Re^m\to\Re$ for equality constrained problem given by

$$\psi(x,\lambda;r)=f(x)+\lambda^T h(x)+\frac{r}{2}\|h(x)\|^2, \tag{1.18}$$

where $r>0$ is the penalty parameter.

There are two mechanisms by which unconstrained minimization of $\psi(x,\lambda;r)$ can yield points close to an optimal solution x_* to equality constrained optimization problem.

a) By taking λ close to λ_*. It is shown by Bertsekas [14] that, if λ is sufficiently close to λ_*, and r is greater than some threshold value, then the local minimization of $\psi(x,\lambda;r)$ is a very good approximation of x_*. The difficulty is that we don't know λ_*.

b) By taking r very large. If there is very high cost for infeasibility, the unconstrained minimum of $\psi(x,\lambda;r)$ will be nearly feasible. This means that $\psi(x,\lambda;r)\approx f(x)$ for nearly feasible x $(i.e., h(x)\approx 0)$. Hence, we can expect to get a good approximation of x_* by unconstrained minimization of $\psi(x,\lambda;r)$ where r is large.

The augmented Lagrangian method is motivated by the preceding considerations. It consists of solving a sequence of problems of the form

$$\text{Minimize } \psi(x,\lambda;r) \tag{1.19}$$

$$\text{subject to } x\in\Re^n; \tag{1.20}$$

where $\{\lambda_k\}$ is a sequence in $\Re^m$ and $\{r_k\}$ is sequence of penalty parameters satisfies $0<r_k<r_{k+1}$ for all k.

The first update formula for the Lagrange multiplier λ was suggested by More [19] and Philip [20]. It is given by

$$\lambda_{k+1} = \lambda_k + r_k h(x_k). \tag{1.21}$$

D) Sequential Quadratic Programming (SQP) Method

SQP method is a direct technique. It is one of the most effective methods and very successful for solving constrained optimization problem. The theory was initiated by Wilson [21], and was further developed by Han [22, 23], Powell [24], and Schittkowski [25, 26].

SQP method generates steps by solving quadratic programming subproblems. Some SQP methods employ convex quadratic programming subproblms for the step computation (typically using quasi-Newton Hessian approximations) while other variants define the Hessian of the SQP model using second derivative information, which can lead to nonconvex quadratic programming subproblems.

SQP method is appropriate for small or large problem. Also, SQP methods show their strength when solving problems with strength nonlinear constraints. The primary goal of these algorithms is to find a point that satisfies the first order necessary optimality conditions.

SQP method is one of the most effective method for solving equality constrained optimization problem. It generates steps by solving quadratic programming subproblems. The quadratic programming subproblem consists of minimizing a quadratic approximation to the Lagrangian function of the original optimization problem subject to linear approximation of the constraints, as follow

$$\text{Minimize} \quad \nabla_x L(x_k, \lambda_k)^T d + \frac{1}{2} d^T H_k d \tag{1.22}$$

$$\text{subject to} \quad \nabla h(x_k)^T d + h(x_k) = 0, \tag{1.23}$$

where H_k is the Hessian of Lagrangian function or an approximation to it. At each iteration, the above quadratic programming subproblem is solved. Suppose that the point x_k is regular, $H_k = \nabla_{xx}^2 L(x_k, \lambda_k)$, and $\nabla_{xx}^2 L(x_k, \lambda_k)$ is positive definite on the null space of $\nabla h(x_k)^T$. Then the solution d_k and the corresponding multipliers $\Delta\lambda_k$ of the quadratic programming subproblem are equal to Newton step on the system of first order necessary optimality conditions

$$\begin{aligned} &\nabla f(x) + \nabla h(x)^T \lambda = 0 \\ &h(x) = 0, \end{aligned}$$

Given by the solution of the linear system

$$\begin{bmatrix} \nabla_{xx}^2 L(x_k, \lambda_k) & \nabla h(x_k) \\ \nabla h(x_k)^T & 0 \end{bmatrix} \begin{bmatrix} d \\ \Delta\lambda \end{bmatrix} = \begin{bmatrix} -\nabla_x L(x_k, \lambda_k) \\ -h(x_k) \end{bmatrix}$$

An extensive survey on SQP algorithms is given by Lalee, Nocedal, and Plantenga [27], Murray and Prieto [28], Yuan and Sun [29], and Nocedal and Wright [30].

- ***Difficulties and limitations of classical optimization techniques***

The classical techniques are usually slow, requiring many function evaluations for convergence and are not efficient in nondifferentiable or discontinuous problems [2, 31]. There are some common difficulties with the classical techniques:

a) Convergence to an optimal solution depends on the chosen initial solution.
b) Most algorithms are prone to get stuck to a suboptimal solution.
c) An algorithm efficient in solving one problem and may not be efficient in solving a different problem.
d) Algorithms are not efficient in handling problems having discrete variables or highly nonlinear and many constraints.
e) Algorithms cannot be efficiently used on a parallel computer.

1.4.2 Advanced Techniques

The advanced techniques overcome difficulties and limitations of classical techniques. These modern techniques are used to solve linear, nonlinear, differential and non-differential optimization problems. They don't require chosen initial solution. The advanced techniques are less susceptible to getting 'stuck' at local optimal. It requires only the function values (and not the derivatives). They are efficient in handling problems having discrete variables or highly nonlinear and many constraints. Some more popular advanced optimization techniques are genetic algorithms, simulated annealing, particle swarm optimization, ant colony optimization, artificial bee colony, tabu search, neural network-based optimization and harmony search [6, 7].

1.4.2.1 Genetic algorithm (GA)

GA was first proposed by John Holland (1975) [31]. It is based on the process of Darwin's theory of Evolution. GA starts with a set of potential solutions or chromosomes that are randomly generated or selected. The potential solutions are replaced with new generations during several iterations hoping to converge to the most 'fit' solution. New generations (offspring) are generated by applying GA operators. GA operators are crossover and mutation operators. Crossover includes splitting of two chromosomes and combining one half of each chromosome with the other pair. Mutation includes flipping a single bit of a chromosome. Then, the chromosomes are evaluated using a certain fitness criteria and the best ones are kept for the next generation while the others are discarded. This process is repeated until one chromosome has the best fitness. The best fitness chromosome is taken as the best solution of the problem. A typical flow chart of GA is showed in Figure 1.2.

GA has several advantages. GA advantages as it works well for global optimization especially where the objective function is discontinuous or with several local minima. GA advantages lead to some potential disadvantages. GA disadvantages as it does not use extra information such as gradients and it has a slow convergence rate on well-behaved objective functions.

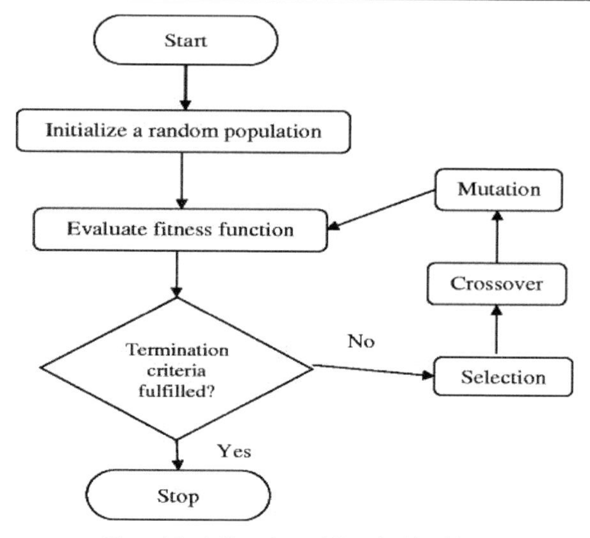

Figure 1.2 A flow chart of Genetic Algorithm

1.4.2.2 Simulated annealing (SA)

SA is described by Kirkpatrick and Daniel et al in 1983 [32]. SA name and inspiration come from the annealing process in metallurgy. SA technique involved heating process and controlled cooling process of a material. Heating process increases the size of its crystals and reduce their defects. The heat causes the atoms to become far from their initial positions (a local minimum of the internal energy) and search randomly through the states of higher energy. The slow cooling gives atoms more chances of finding configurations with lower internal energy than the initial one. In SA technique, each point of the search space is interpreted with a state of physical system. The function to be minimized is interpreted as the internal energy of the system in that state. The goal in SA technique is to bring the system from an arbitrary initial state to a state with the minimum possible energy [32, 33]. A typical flow chart of simulated annealing is showed in Figure 1.3

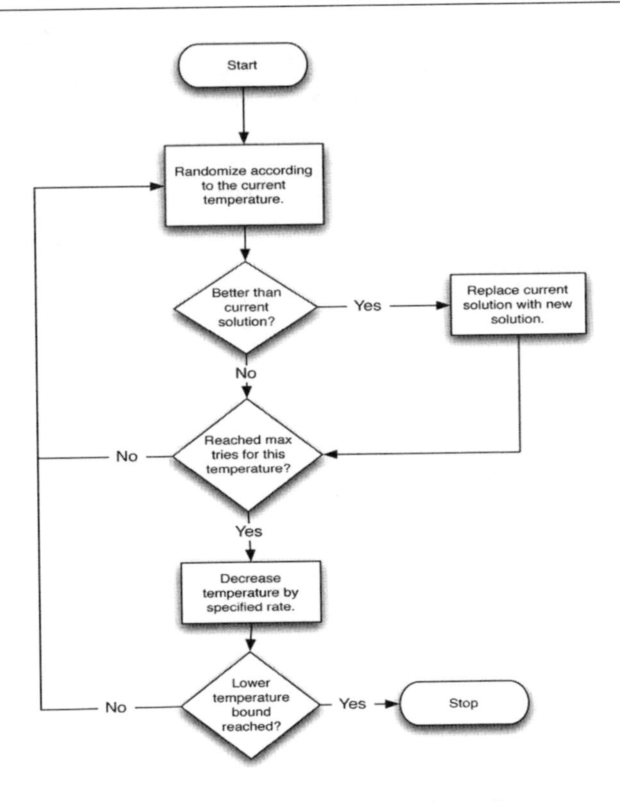

Figure 1.3 A flow chart of Simulated Annealing

1.4.2.3 Neural network optimization

The neural network optimization was originally used by Hopfield and Tank in 1985 [34]. It consists of a number of sub-units called neurons. These are interconnected in parallel to form a network typical neuron consists of inputs, a summing function, a limiting or threshold function and outputs. These correspond to the dendrons, the synapses, the cell bodies and the axons of a biological neuron. For neural networks, to perform any useful function, they must be trained. The most common method of training has been found the back propagation method. In the back propagation method, an error function is minimized by changing the weights of each synapse in proportion to the error calculated. The error is the difference between some target or desired value and the actual value (output) obtained. The analogy between the error function in back propagation training and the objective function in optimization is too obvious to be missed. Hence, the application of neural nets to optimization problems is easy to expect [34, 35]. A typical flow chart of neural network optimization is showed in Figure 1.4.

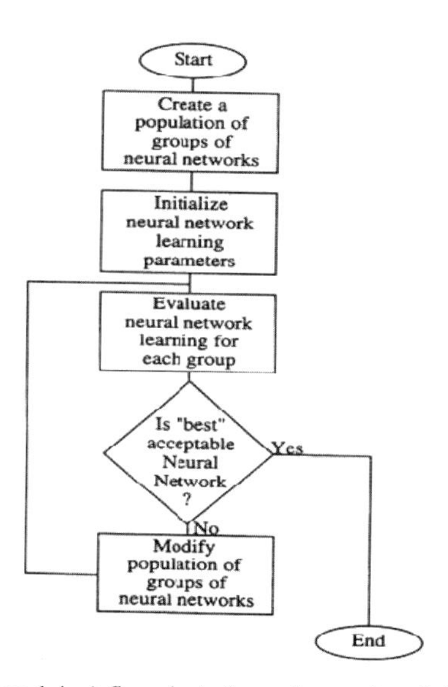

Figure 1.4 A flow chart of neural network optimization

1.4.2.4 Tabu search (TS)

TS is a kind of neighborhood search. It has been mainly propagated by Fred Glover at 1986 [36]. TS starts with a random solution and then neighborhood of current solution is searched. The change in solution value from one iteration to other is called as a move. As the current solution moves, in each iteration, the neighborhood also changes. The possible neighborhood search is restricted by the number of solutions and the constraints of the optimization problem. The restrictions on the movement of a solution are also placed due to the Tabu list. Tabu list contains a set of moves which are prohibited. The list generally is used to prevent cycling and it removes the chances of searching a solution which has been previously visited. The size of tabu list plays an important role in optimization of problem. The tabu list keeps changing as the search process progresses. The older entries in the tabu list are eliminated with newer entries. An aspiration criterion may also be introduced to override the tabu list, which helps to achieve better results. The search process can be stopped after some predefined number of iterations or if the optimum result does not change for a certain number of iterations [36, 37].

1.4.2.5 Ant colony optimization (ACO)

ACO is a nontraditional technique. It is proposed by Dorigo in 1992 [39]. ACO is inspired from the behavior of ants (e.g., a trace of a chemical substance that can be smelled by other ants trails in search of food). ACO belongs to a class of algorithms which can be used to obtain good enough solutions in reasonable computational time for combinatory optimization problems. Ants communicate with one another by depositing pheromones. Initially in search of food, ants wander randomly and upon finding a food source, return to their colony. On their way back to the colony, they deposit pheromones on the trail. Other ants then tend to follow this pheromone trail to the food source and on their way back may either take a new trail, which might be shorter or longer than the previous trail, or would come back along the previous laid pheromone trail. Also, on their way back, the other ants deposit pheromones on the trail. Pheromones have a tendency to evaporate with time. Hence, over a period of time, the shortest trail (path) from the food source to the colony would become more attractive and have a larger amount of pheromone deposited as compared with other trails. An explaining of the above defined steps is shown in flow chart in Figure 1.5 below.

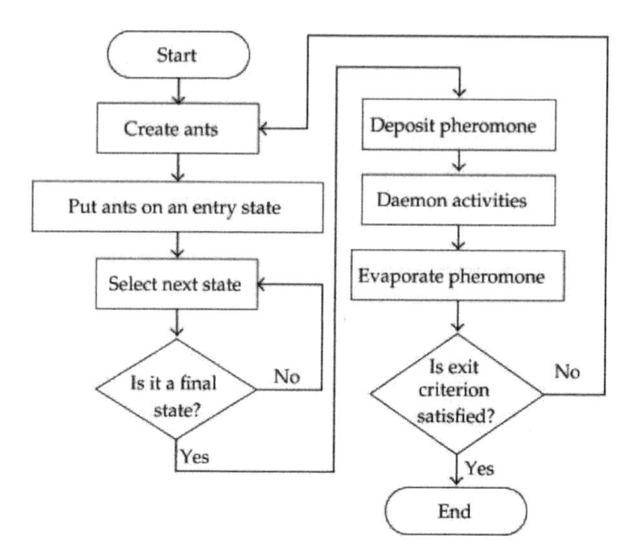

Figure 1.5 A flow chart of Ant colony Optimization

1.4.2.6 Particle swarm optimization (PSO)

PSO is an optimization technique based on the movement of intelligence swarms. It was introduced in 1995 [40] by James Kennedy (social psychologist) and Russell Eberhart (Electrical Engineer). PSO is a method for optimizing hard numerical functions on metaphor of fish. Suppose the following scenario, a flock of birds is randomly searching for food in an area, where there is only one piece of food available and none of them knows where it is, but they can estimate how for it would be at each iteration. The problem here what is the best strategy to find and get the food? Obviously the simplest strategy is to follow the bird known as the nearest one to the food. In a PSO algorithm, swarm is initiated randomly with finding the personal best (best value of each individual so far) and global best (best particle in the whole swarm). Initially, each individual with its dimensions and fitness value is assigned to its person best. The best individual among particle best swarm, with its dimension and fitness value is, on the other hand, assigned to the global best. Then a loop starts to converge to an optimum solution. In the loop, particle and global bests are determined to update the velocity first. Then the current position of each particle is updated with the current velocity. Evaluation is again performed to complete the fitness of particles in the swarm. This loop is terminated with a stopping criterion predetermined in advance [40, 41]. A typical flow chart of PSO is showed in Figure 1.6.

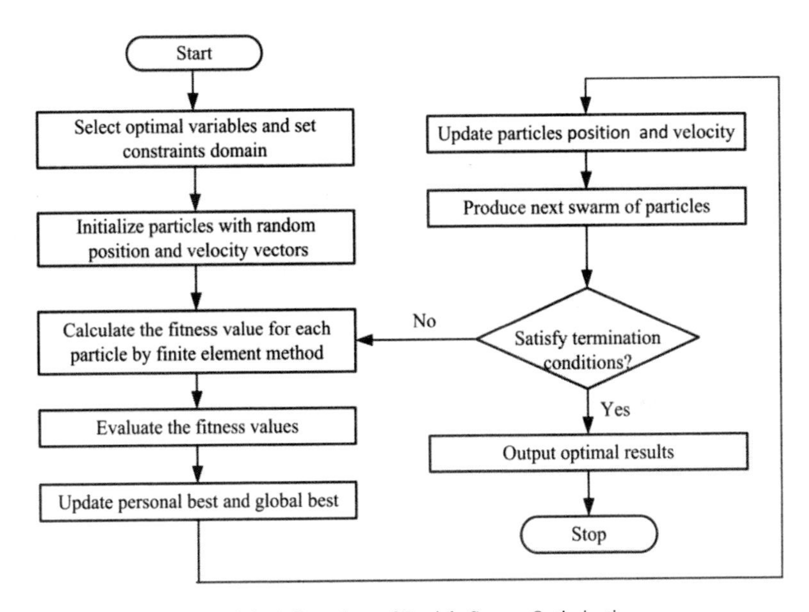

Figure 1.6 A flow chart of Particle Swarm Optimization

1.4.2.7 Harmony search (HS)

Recently, Geem & Lee developed HS meta-heuristic algorithm at 2005 [41]. HS algorithm is based on natural musical performance processes that occur when musicians search for a better state of harmony, such as during jazz improvisation. As compared to conventional mathematical optimization algorithms, the HS algorithm imposes fewer mathematical requirements to solve optimization problems, and the probability of becoming entrapped in a local optimum is reduced in this algorithm. Since the HS algorithm uses a stochastic random search, it has a new-paradigmed derivative [41]. The algorithm considers several solution vectors simultaneously, in a manner similar to the GA

1.4.2.8 Artificial bee colony (ABC)

Recently, Karaboga developed a new optimization algorithm called the ABC at 2005 [42]. Bee algorithms form another class of algorithms which are closely related to the ant colony optimization. Bee algorithms are inspired by the foraging behavior of honey bees. Honey bees live in a colony and they forage and store honey in their constructed colony. Honey bees can communicate by pheromone and 'waggle dance'. For example, an alarming bee may release a chemical message (pheromone) to stimulate an attack response in other bees. Furthermore, when bees find a good food source and bring some nectar back to the hive, they will communicate the location of the food source by performing the so-called waggle dances as a signal system. Such signaling dances vary from species to species, however, they will try to recruit more bees by using directional dancing with varying strength so as to communicate the direction and distance of the found food resource.

In ABC model, the foraging bees are classified into three different types: employed bees, onlookers and scouts. A bee which has found a food source to exploit is called an employed bee. Onlookers are those waiting in the hive to receive the information about the food sources from the employed bees and Scouts are the bees which are randomly searching for the new sources of food around the hive. Employed bees go to their food source and return to hive and dance on this area. Then, the employed bee whose food source has been abandoned becomes a scout and starts to search for finding a new food source. Onlookers see the dances of employed bees and choose food sources depending on the type of dances [41, 42]. A typical flow chart of ABC is showed in Figure 1.7.

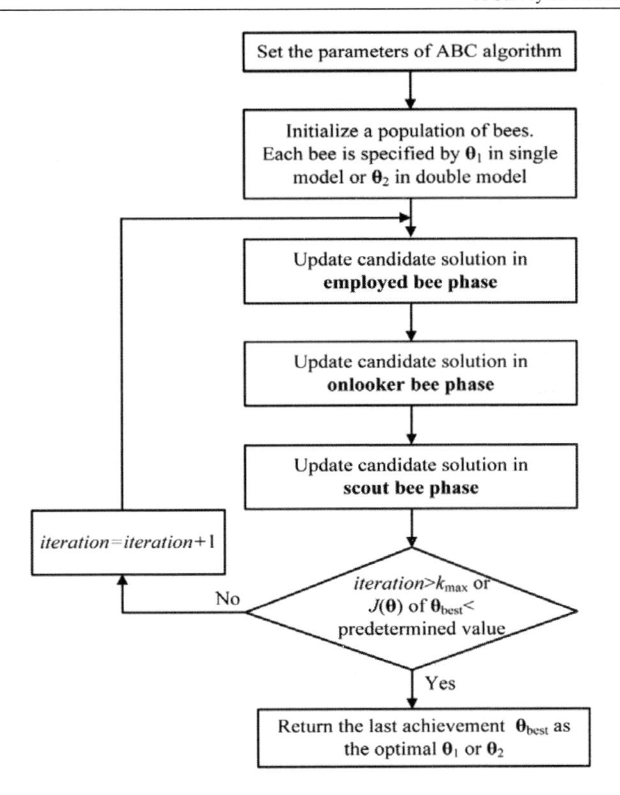

Figure 1.7 A flow chart of Artificial Bee Colony

CHAPTER 2

Genetic Algorithm

2.1 Introduction

GA is represented an efficient global method for optimization problems especially for nonlinear optimization problems. Many practical optimum design problems are not standard optimization algorithms. The objective function may be discontinuous, stochastic, highly nonlinear or haven't derivate. If the classical optimization techniques are used for this type of problem they will be inefficient, computationally expensive and in most cases, find a relative optimum that is closest to the starting point. GA is well suited for solving such problems. They do not require linearization assumptions or calculation of partial derivatives. The additional advantage over classical methods is that the sampling is global, rather than local [43, 44].

GA is stochastic search technique that is based on the mechanism of natural selection and natural genetics. GA differs from classical search techniques. It starts by generating an initial set of random solutions called population satisfying boundary constraints of the problem. Each individual in the population is called individual (or a chromosome). First, the initial population is formed by random choosing of set of chromosomes from the search space. Next, new generation of chromosomes is generated by applying the genetic search operators one after another. The expected quality over all the chromosomes of new generation is better than that of the previous generation [31, 43]. Generating new generation is repeated to have best chromosomes. The best chromosomes of the last generation are reported as the final.

The most important aim of this chapter is to introduce the working principles of GA, GA procedure for solving optimization problems, GA parameters that effect on GA performance. In final we introduce advantages and disadvantages of GA.

2.2 Working Principles of GA

The GA workability is based on Darwinian's theory of survival of the fittest. GA may contain a chromosome, a gene, population, fitness function, selection, crossover, and mutation. GA starts with a set of potential solutions represented by chromosomes called initial population. Initial population is used to form a new population. The initial population is replaced with new population, that is, offsprings. The new population during several iterations hopes to converge on the most 'fit'. New populations are generated by applying GA operators. GA operators are crossover and mutation operators. This process is repeated until some condition is satisfied [31, 44]. Algorithmically, the basic GA is outlined as below:

Step 1: Start GA by generation of random population of chromosomes that satisfy boundary constraints to the problem.

Step2: Evaluate the fitness of chromosomes in the population.

Step3: Create a new population by repeating the following steps until the new population is complete applying the following steps:

a) According to their fitness, select two parent chromosomes from a population. The chromosomes with better fitness have the bigger chance to be selected as a parent.

b) By the crossover probability, crossover the parents to generate new offspring (child).

c) By the mutation probability, mutate some parents to generate new offspring.

d) Merge the old population with new offspring to form the new population.

Step 4: If the end condition is satisfied, stop, and return the best solution in current population otherwise new generated population is used for a further run of the algorithm then go to step2.

2.3 Genetic algorithm procedure for optimization problems

The first decision in applying GA, to seek optimal values for vari1ables, is how to represent design parameters of an individual. The variables are represented one of encoding techniques representation. In encoding techniques, genes are represented using bits, arrays, numbers, trees or any other objects. Encoding technique depend on the problem. Here, we introduce some encodings techniques, which have been already used with some success in such problems.

2.3.1 Encoding

GA works on two types of spaces alternatively, coding space (genotype) and solution space (phenotype). The phenotype describes the outward appearance of an individual. A transformation exists between genotype and phenotype, also called mapping, which uses the genotypic information to construct the phenotype. A chromosome refers to a string of certain length where all the genetic information of an individual is stored. Each chromosome consists of many gens. Gens are the smallest information units in a chromosome. As shown in Figure 2.1, GA search over the genotype world and solutions which are obtained by GA are evaluated and selected into phenotype world [45, 46].

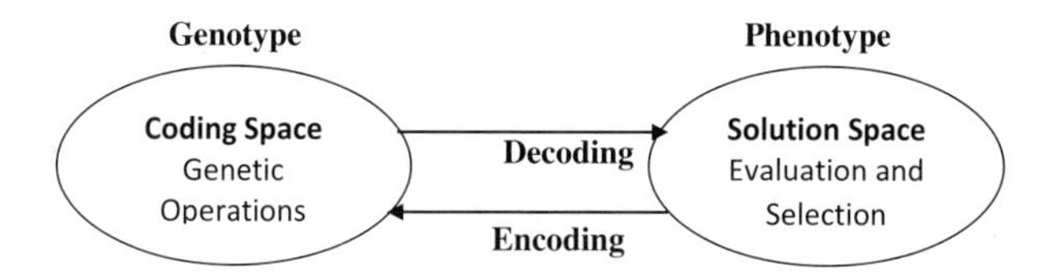

Figure 2.1: Encoding – Decoding method

A) Real-valued encoding

In real-valued encoding an individual is characterized by a vector of real numbers. It is more natural to use the floating-point representation for real parameter optimization problems because it is closest to the real design space, and moreover, the string length is reduced to the number of design variables.

B) Binary encoding

Binary encoding is the most common form of encoding because of its relative simplicity and that the first research of GA used this type of encoding. It is used when fitness depends on both value and order as Knapsack problem. In Knapsack problem, we have things with given value and size and knapsack has given capacity. We need to select the things to maximize the value of things in knapsack, but do not extend knapsack capacity. In binary encoding the value of data is converted into binary strings [45]. Figure 2 .1 illustrates chromosome represented in binary encoding.

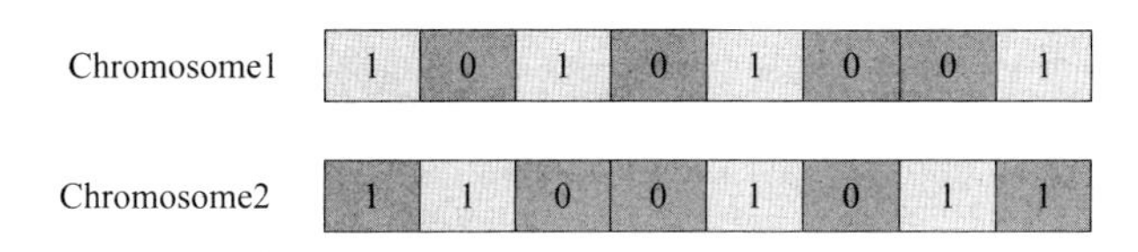

Figure 2.2 Binary encoding

- **Mathematical formulas for the binary encoding and decoding**

The mathematical formulas for the binary encoding of nth variable p_n are given as follows [31]:

$$p_{norm} = \frac{p_n - p_{lo}}{p_{hi} - p_{lo}} \tag{2.1}$$

$$gene[m] = round\{p_{norm} - 2^{-m} - \sum_{p=1}^{m-1} gene[p].2^{-p}\} \tag{2.2}$$

The mathematical formulas for the binary decoding of nth variable, are given as follows:

$$p_{quant} = \sum_{m=1}^{N_{gene}} gen[m].2^{-m} + 2^{-(m+1)} \tag{2.3}$$

$$q_n = p_{quant}(p_{hi} - p_{lo}) + p_{lo}; \tag{2.4}$$

where,

p_{norm} : normalized variable, $0 \le p_{norm} \le 1$

p_{lo} : smallest variable value

p_{hi} : highest variable value

$gen[m]$: binary version of p_n

$round\{.\}$: round to nearest integer

p_{quant} : quantized version of p_{norm}

q_n : quantized version of p_n

C) Permutation encoding

Permutation encoding is suitable for ordering or queuing problems as travelling salesman problem [46]. Every chromosome in permutation encoding is a string of numbers in a sequence. Figure 2.3 illustrate chromosome represented in permutation encoding

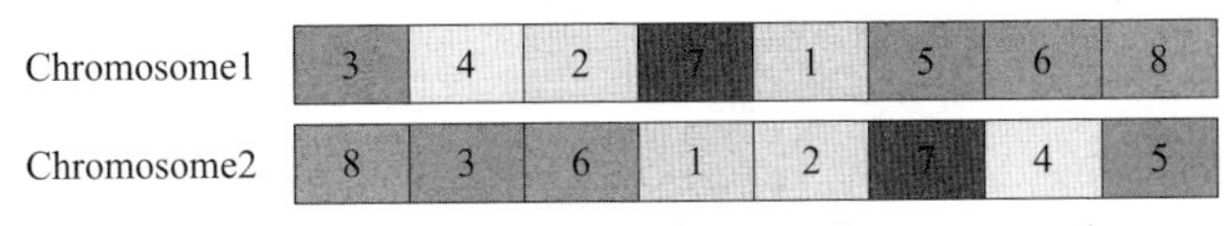

Figure 2.3 Permutation encoding

D) Value encoding

In value encoding technique, every chromosome is a string of some values connected to the problem such as number, characters or objects. Value encoding is used where some more complicated values are required [45, 46]. Using binary will be difficult in this type. Figure 2.4 illustrate chromosome represented in value encoding.

Chromosome 1	2.4351	3.8609	4.110	6.783
Chromosome 2	ABDJE	IFJDH	DIERJ	LFEGT
Chromosome 3	Right	Back	Forward	Left

Figure 2.4 Value encoding

E) Tree Encoding

Tree encoding technique is suitable for evolving programs or expressions for genetic programming. Every chromosome in tree encoding is a tree of some functions or commands in programming languages [45, 46]. Tree encoding can be expressed as shown in Figure 2.5.

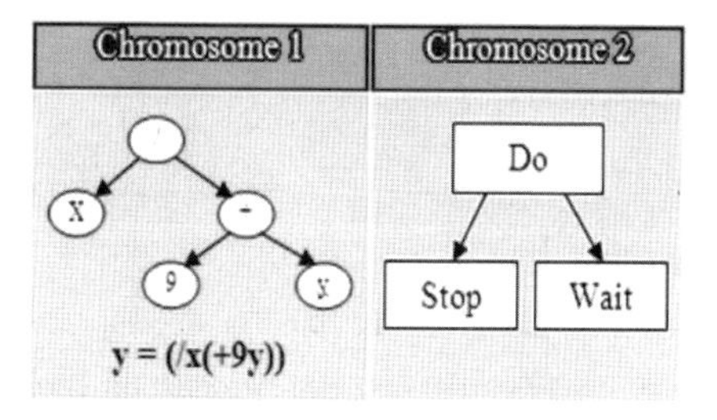

Figure 2.5 Tree Encoding

2.3.2 Initial Population

The GA starts with a group of chromosomes known as the population. The gens in every chromosome have to be generated randomly within specified boundaries. Here we propose the generation of initial population with real-valued representation and with one of encoding techniques (binary encoding).

- **Initial Population for real-valued encoding**

The initial population is presented as a matrix named as " pop ". Number of rows of matrix" pop ", named a " N_{chrom} ", is the number of generated chromosomes. N_{chrom} depends on the nature of the problem, but typically contains several hundreds or thousands of possible solution. Number of columns of matrix" pop ", named as " N_{var} ". is number of parameters (variables). Given an initial population, the full matrix (pop) has size $(N_{chrom} * N_{var})$. All genes are generated within specified boundaries using the formula [31]:

$$p = (p_{hi} - p_{lo}) * Rand + p_{lo} \tag{2.5}$$

where,

p_{hi} : Highest number in the variable range

p_{lo} : Lowest number in the variable range

$Rand$: Random valve between 0 and 1.

- **Initial Population for binary encoding**

The population is an (N_{pop}, N_{bits}) matrix filled with random ones and zeros generated and has N_{pop} chromosomes. The Initial Population is generated using following equations:

$$m = \log_2(p_{\max} + 1) \tag{2.6}$$

$$N_{bits} = (m * N) \tag{2.7}$$

$$pop = round(rand(N_{pop}, N_{bits})); \tag{2.8}$$

where,

m : number of bits for every variable

N : number of variables .

$p_{\max}$: the highest value of variables boundaries.

The function(N_{pop}, N_{bits}) generates a $N_{pop} * N_{bits}$ matrix of uniform random numbers between zero and one. The function rounds the numbers to the closest integer which in this case is either 0 or 1. Each row in the pop matrix is a chromosome [31].

2.3.3 Evaluation

The first step after creating a generation is to calculate the fitness value of each member in initial population. The fitness of all individuals is evaluated using objective function values. Chromosomes, which are more optimal, are allowed to breed and mix their datasets by any of several techniques to produce a new generation that will (hopefully) be even better. For the real-valued encoding, the fitness of all individuals is evaluated by direct substitution with chromosomes into objective function values. For other encoding techniques [45], the evaluating process of a chromosome consists of these two steps:

1) Conversion the chromosome from its genotype to its phenotype.
2) Evaluation the chromosome using the objective function.

2.3.4 Create a new population

After evaluation, a new population of chromosomes is obtained by applying the genetic search operators one after another. The genetic search operators are selection, mutation and crossover. The expected quality of a new population over all the chromosomes is better than that of the current population.

2.3.4.1 Selection

Selection is an important operator in GA, based on an evaluation criterion. The chromosomes which are more optimal are selected and allowed to breed and mix to produce a new generation that will hopefully be even better. In selection stage, individual are chosen from the string of chromosomes. There is no difference for selection process applied with the real-valued representation and for encoding techniques representation. The selection treated with fitness which is already real values. The commonly used techniques for selection of chromosomes are roulette wheel, stochastic universal sampling, rank selection, tournament selection, and steady state selection [48, 49].

A) Roulette wheel s election

In roulette wheel selection, the parents are selected according to their fitness. It is the most common method for implementing fitness with proportionate selection. Each individual is assigned a slice of circular roulette wheel. The size of slice is proportional to the individual fitness of chromosomes. The larger size of slice has the bigger chance to be selected [48]. The roulette wheel algorithm is described as following:

Step 1: Calculate fitness values of every individual in the population.

Step 2: Calculate every individual probability by dividing individual chromosome's fitness by the sum of fitness values of whole population. The probability of selection of an individual:

$$p(a_i) = \frac{f(a_i)}{\sum_{i=1}^{n} f(a_j)} \qquad , j = 1, 2, \ldots, n; \tag{2.9}$$

where $'n'$ is the size of population, $f(a_i)$ is the fitness value of the individual a_i .

Step 3: Partition the roulette wheel into sectors according to the calculated probabilities in step 2.

Step 4: Spin the wheel $'n'$ number of times. When the roulette stops, the sector on which the pointer points corresponds to the individual being selected. Figure 2.6 shows roulette wheel or five individuals having different fitness values. At spinning the wheel, the pointer more stops in section of the individual with a larger size of slice in the wheel. From the figure, the individual with number three has bigger chance to be selected.

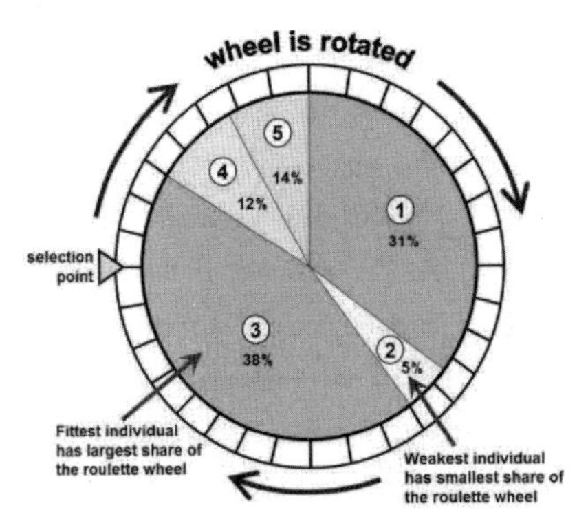

Figure 2.6 Roulette wheel selection

B) Stochastic Universal Sampling (SUS):

Stochastic universal selection was developed by Baker at 1987 [49]. SUS selection is started by sizing the slots of a weighted as in roulette wheel selection technique. Instead of a single selection pointer employed in roulette wheel methods, SUS uses N equally spaced pointers, where $'N'$ is the number of selections required (population size). The position of the first pointer is given by a randomly generated number in the range.

After that, the wheel is spun just once and the $'N'$ individuals that situated in front of the determined points are chosen. SUS selection ensures a selection of offspring, which is closer to what is deserved than roulette wheel selection. SUS selection can be used to make any number of selections [48, 49].

C) Rank selection method

When the fitness value of chromosomes differs very much, applying roulette wheel selection techniques is not satisfactory in GA. Often, the individual with largest fitness will be selected and other individuals will have minimum chances to be selected. Rank selection method is a slower convergence technique. First, rank selection sorts the population according to fitness value and ranks them. Rank N is assigned to the best individual and rank 1 to the worst individual. Then every chromosome n is allocated selection probability with respect to its rank. Individuals are selected as per their selection probability. This method prevents quick convergence and the individuals in a population are ranked according to the fitness and the expected value of each individual depends on its rank rather than its absolute fitness. The rank selection method is shown in Figure 2.7. For example, if the best individual fitness is 80 percent, its circumference occupies 80 percent of the roulette wheel as in Figure 2.7 (A) and then other individuals will have minimum chances to be selected. On the other hand as in Figure 2.7 (B), the rank selection first ranks the population according to their fitness. The circumference of best individual occupies 45 percent of the wheel [49].

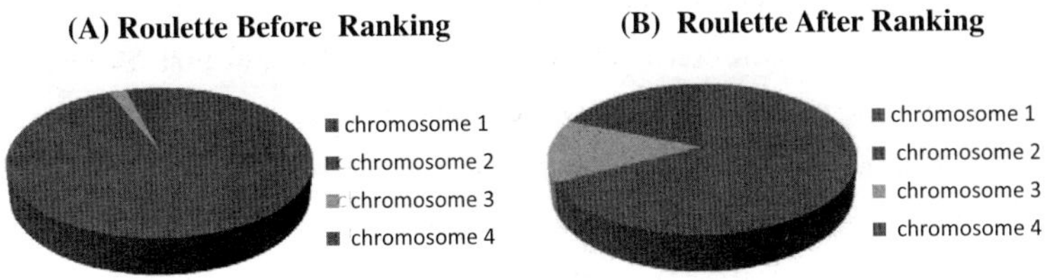

(A) Roulette Before Ranking

chromosome 1
chromosome 2
chromosome 3
chromosome 4

(B) Roulette After Ranking

chromosome 1
chromosome 2
chromosome 3
chromosome 4

Figure 2.7 Rank selection

D) Tournament Selection

Tournament selection was presented by Goldberg at 1989 [49]. Tournament selection started with selecting $'t'$ number of individuals. The $'t'$ individuals are randomly selected from the population. Tournament selection selects the best individual from this group (tournament) as parent. The best individual from the tournament is the one with the highest fitness, among $'t'$ individuals. The tournament competition is repeated. Increasing the tournament size, $'t'$ will increase selection pressure. The winner from a larger tournament, on an average, has a higher fitness than the winner of a smaller tournament. Tournament selection is not very useful if used population is large because we will need a lot of time [49].

E) Steady-state selection

In steady-state selection, there is not particular method of selecting parents. Main idea of steady-state selection is that big part of individuals should remain to the next generation. GA works in a following way. In every generation, some good individuals (with high fitness) are selected for creating a new offspring. Then some bad individuals (with low fitness) are removed and the new offspring is placed in their place. The rest of population remains to new generation [48, 49].

2.3.4.2 Crossover

Crossover is one of main genetic operators. It works on two chromosomes at a time and generates offspring by combining both chromosomes' characters. New strings (offspring) are produced by exchanging segments from the parents' strings. The simple way to start crossover would be to choose a random cut-point. Then segment of one parent to the left of the cut-point and segment of the other parent to the right of the cut-point are combined together to generate the offspring. There are different types of crossover and the choice of the crossover type is strictly depends upon the problem. Here, we introduce different types of crossover proposing some of them for different encoding techniques [50]. Types of crossover are:

A) Single Point Crossover

Single point crossover is widely used crossover type and it is the most popular crossover type. Single point crossover chooses a random cut-point. Then bits of one parent to the left of the cut-point and bits of the other parent to the right of the cut-point are combined together to generate the offspring. The better children can be obtained by choosing suitable site of cut-point else it harshly hamper string quality.

B) N-Point Crossover

The N-point crossover was presented by De Jong in 1975 [60]. It has many cut-points sites but it has same rule used in single point crossover. In two-point crossover, there are two sites for cutting. Adding more and more cut-points sites effect the disruptions of building blocks of offspring that sometimes reduce algorithm performance

C) Uniform Crossover

Uniform crossover does not part the chromosomes to be recombined. There are binary crossover mask of same length as the length of the parent chromosomes. This crossover mask is randomly generated for each pair of parent chromosomes. In uniform crossover, each gene in offspring is formed by copying from one of two parents. This parent is chosen according to the corresponding bit in the crossover mask. If the bit in crossover mask is 1, then the resultant gene is copied from the first parent and if the bit in crossover mask is 0, then the resultant gene is copied from the second parent. So, the offspring will have a mixture of genes from both the parents.

D) Three Parents Crossover

Three parent crossover starts by arbitrarily chose of three parents. If there are two genes are similar in the first two parents, the gene is occupied for offspring or else the equivalent gene from the third parent. This type of crossover is mostly used in case of binary encoded chromosomes.

E) Arithmetic Crossover

This type of crossover is used for real-value encoding. Arithmetic crossover operators using linear combine of two parent chromosomes. In arithmetic crossover, two chromosomes are particular randomly mixed to generate the new offspring.

F) Partially mapped Crossover

Partially Mapped Crossover (PMX) is the most frequently used crossover operator for permutation encoding. PMX was proposed by Goldberg and Lingle [61] for travelling salesman problem. PMX tends to respect the absolute positions of genes. In PMX, two chromosomes are associated and two crossover sites are chosen arbitrarily. It copies the portion of chromosome elements between the two crossover points directly to the offspring from one chromosome. The remained genes are inserted through position-by position exchange operations.

G) Order Crossover (OX)

OX was proposed by Davis [50] and also used for chromosomes with permutation encoding. OX tends to respect the relative positions of genes. The process starts by choosing OX applies sliding motion to fill up the left out holes. It copies the portion of chromosome elements between the two crossover points directly to the offspring from one chromosome. The remained genes are inserted with the same absolute position from second chromosome.

H) Cycle Crossover (CX)

CX is also used for chromosomes with permutation encoding. During recombination in cyclic crossover, there is a limitation that each gene either comes from the one parent or the other. The fundamental model of CX is that each moved gene is jointed with its position. The cycle of gens from parent1is started with the first gene of parent one. Then we look at the gene at the equal position in parent two and go to the position with the same gene in parent one. These steps are repeated until we reached to the first allele of parent one. The genes that formed this cycle is called cycle elements. To produce the first offspring, we copy the cycle elements from parent one and complete this offspring from parent two. Repeat similar steps to produce second offspring. By this way, we have two offspring.

I) Real-valued encoding crossover

Here, we will propose some crossover techniques with real valued encoding :

A) Single Point Crossover

In single point crossover, single point is randomly chosen. The data before the crossover point are exactly copied from first parent and the data after this point are exactly copied from the second parent to create new offsprings [45]. Figure2.8. illustrates the single point crossover.

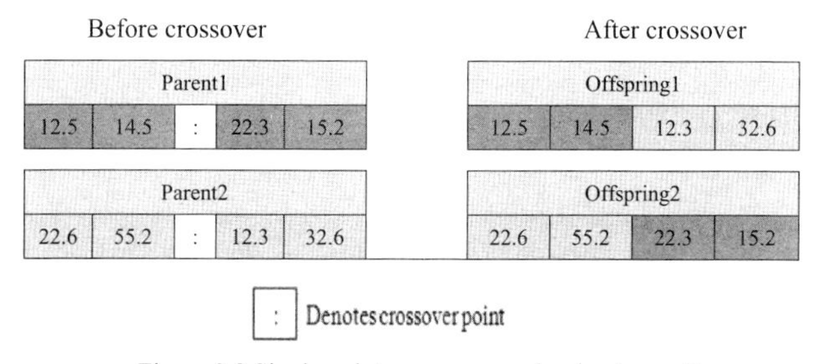

Figure 2.8 Single point crossover, real-valued encoding

B) Two Point Crossover

In this crossover type, two crossover points are chosen. The data between two points are copied from second parent and the data outsides two points are copied from the first parent [54]. The two point crossover is illustrated in Figure 2.9.

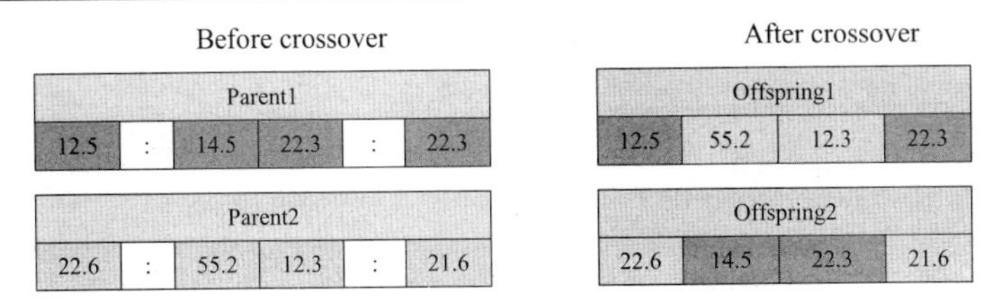

Figure 2.9 Two point crossover, real-valued encoding

C) Uniform Crossover

In uniform crossover, data are randomly copied from first parent chromosome and second parent chromosome [55]. Figure 2.10 illustrated the uniform crossover.

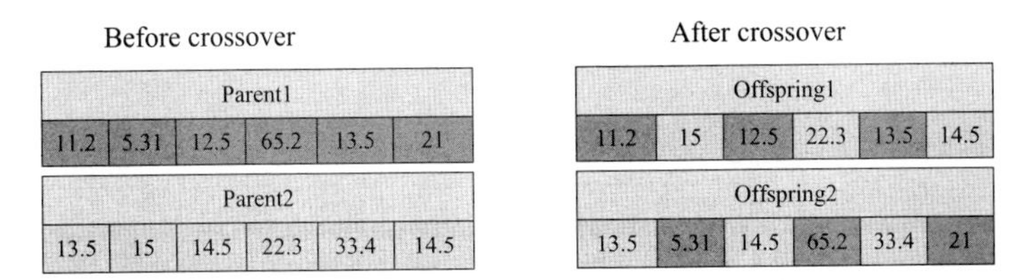

Figure 2.10 Uniform crossover, real-valued encoding

II) Binary encoding crossover

Here, we will propose some crossover techniques with binary encoding [46]:

A) Single Point Crossover

The single point crossover is illustrated in Figure 2.11.

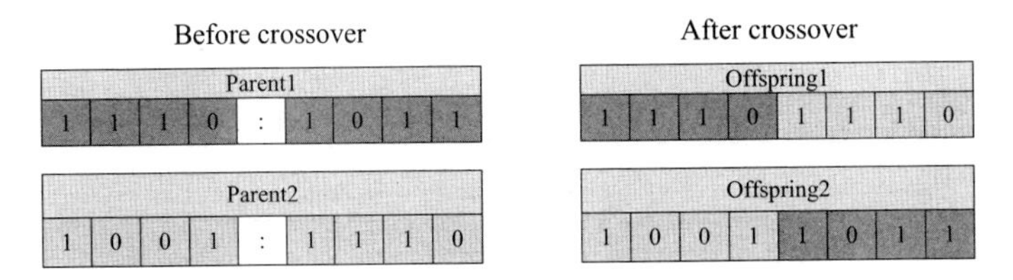

Figure 2.11 Single point crossover, binary encoding

B) Two Point Crossover

The two point crossover is illustrated in Figure 2.12.

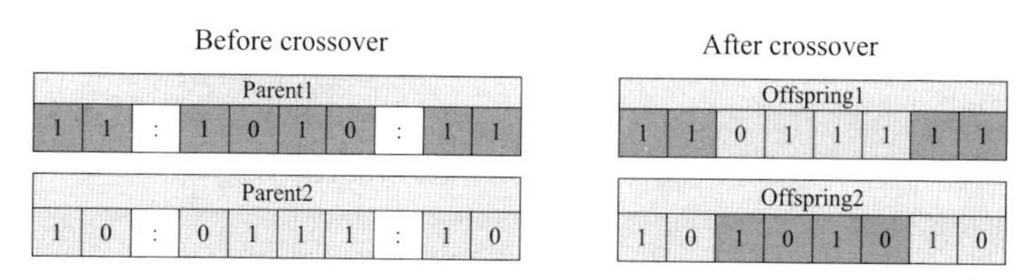

Figure 2.12 Two point crossover, binary encoding

C) Uniform Crossover

Uniform Crossover is illustrated in Figure 2.13.

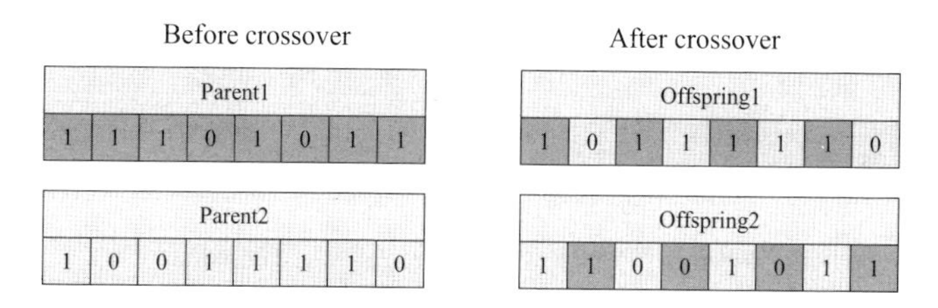

Figure 2.13 Uniform crossover, binary encoding

D) Arithmetic Crossover

In arithmetic crossover, crossover is performed using AND and OR operators. Figure 2.14 illustrated the arithmetic crossover.

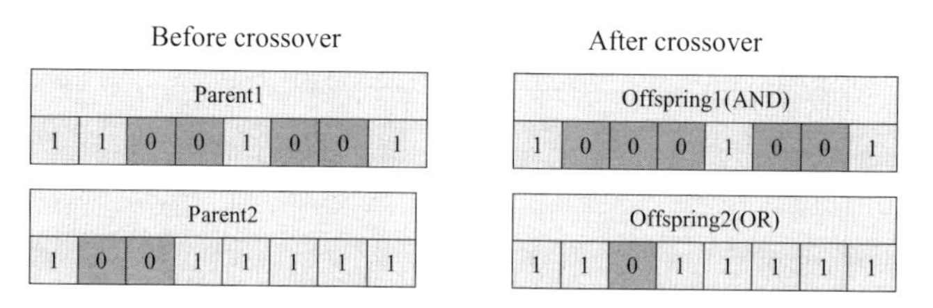

Figure 2.14 Arithmetic crossover, binary encoding

III) Permutation encoding crossover

Here, we will propose single point crossover techniques with permutation encoding:

A) Partial Mapped Crossover

In permutation encoding crossover, partial mapped crossover is mostly used. The PMX starts by selecting a substring from parents at random. Then, these two subgroups are exchanged. The remained genes are inserted through position-by position exchange operations. Figure 2.15 illustrates partial mapped crossover with permutation encoding.

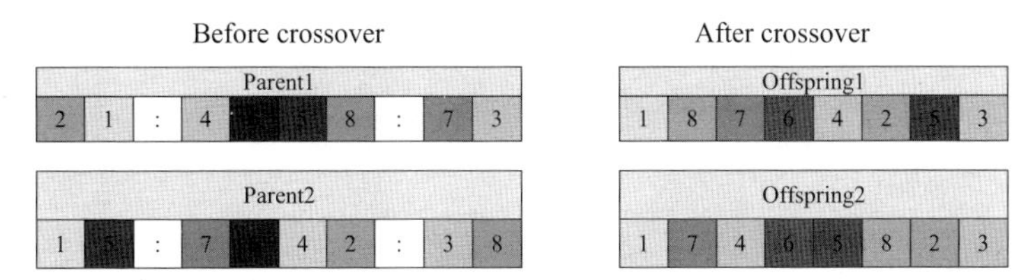

Figure 2.15 Single point crossover, permutation encoding

IV) Value encoding crossover

Here, we will propose single point crossover techniques with tree encoding:

A) Single Point Crossover

It is performed as single point as in binary encoding technique. The single point crossover with value encoding is illustrated in Figure 2.16 [55].

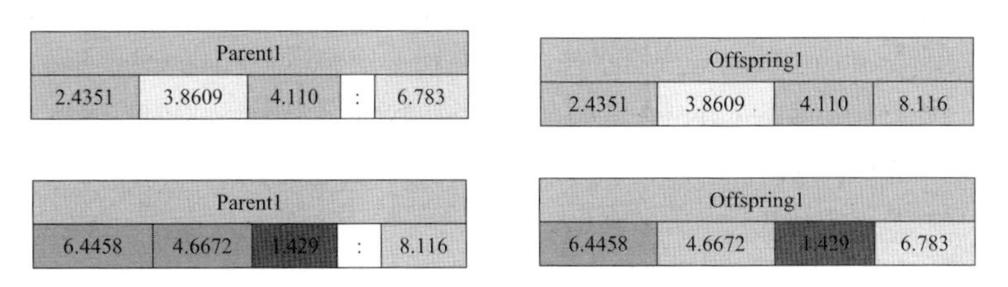

Figure 2.16 Single point crossover, value encoding

V) Tree encoding crossover

Here, we will propose some crossover techniques with tree encoding [54]:

A) Single Point Crossover

In this type of crossover, in both parent tree chromosomes, one point of crossover is selected. Then, the parts of tree below crossover point are exchanged to produce new offsprings. Figure 2.17 illustrated single point crossover with tree encoding.

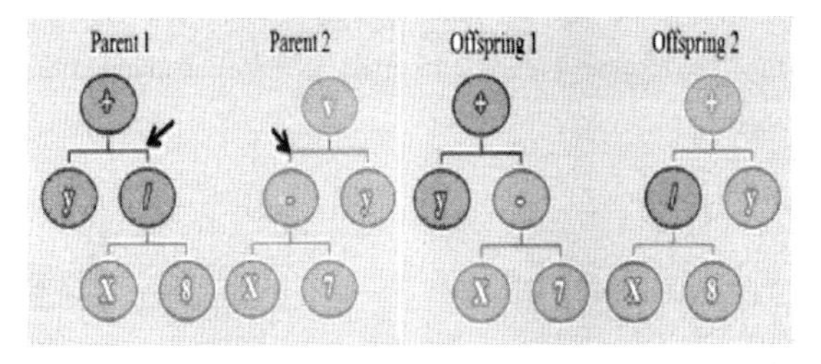

Figure 2.17 Single point crossover, tree encoding crossover

2.3.4.3 Mutation

The critical problem in most optimization techniques is premature convergence. The premature convergence occurs when highly fit parent chromosomes in the population breed many similar offsprings in early evolution time. In this case, crossover operation of GA will not generate quite different offsprings from their parents. Mutation is a common operator used to help preserve diversity in the population by finding new points in the search space to evaluate. When a chromosome is chosen for mutation, a random change is made to the values of some locations in the chromosome. Here, we introduce different types of mutation proposing some of them for real- valued representation and encoding techniques representation. Types of mutation are [62]:

A) Insert Mutation

Insert mutation is used in permutation encoding. First of all, pick two genes values at random. Then move the second gene to follow the first, shifting the rest along to accommodate. Note that this preserves most of the order and the adjacency information [51].

B) Inversion Mutation

Inversion mutation is used for chromosomes with permutation encoding. In order to perform inversion, pick two gens at random and then invert the substring between them. It preserves most adjacency information and only breaks two links but it leads to the disruption of order information.

C) Scramble Mutation

Scramble mutation is also used with permutation encoded chromosome. In this mutation, one has to pick a subset of genes at random and then randomly rearrange the alleles in those positions. Subset does not have to be contiguous.

D) Swap Mutation

It is also used in Permutation encoding. To perform swap mutation select two genes at random and swap their positions. It preserves most of the adjacency information.

E) Flip Mutation

Flip mutation based on a generated mutation chromosome, flipping of a bit involves changing 0 to 1 and 1 to 0. A parent is considered and a mutation chromosome is randomly generated. In Flip mutation, the corresponding bit in parent chromosome is flipped (0 to 1 and 1 to 0) and child chromosome is produced. It is commonly used in binary encoding.

F) Interchanging Mutation

In Interchanging Mutation, two random positions of the string are chosen and the corresponding genes to those positions are interchanged.

G) Reversing Mutation

In case of reversing mutation applied for binary encoded chromosome, random position is chosen and the bits next to that position are reversed and child chromosome is produced.

H) Uniform Mutation

The Mutation operator changes the value of chosen gene with uniform random value. This value is selected between the specified upper and lower bound for that gene. It is used in case of real and integer representation.

I) Creep Mutation

In creep mutation, a random gene is selected and its value is changed with a random value between lower and upper bound. It is used in case of real representation.

I) Real-valued encoding mutation

Here, we will propose interchanging mutation techniques with real valued encoding.

A) Interchanging mutation

Two random positions of the string are chosen and the corresponding genes to those positions are interchanged.

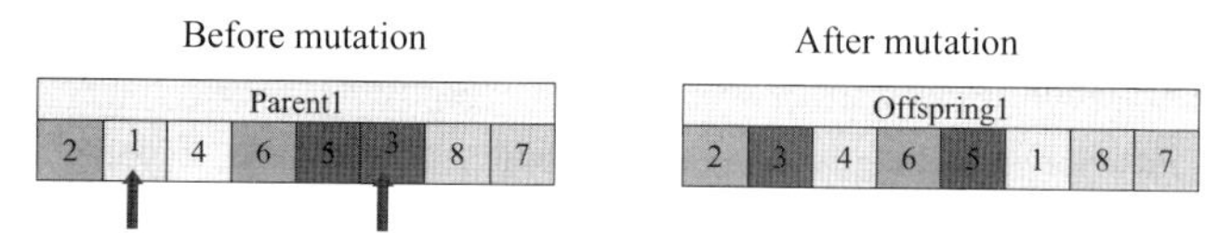

Figure 2.18 Interchanging mutation, real-valued encoding

II) Binary encoding mutation

Here, we will propose flip mutation techniques with binary encoding [46]:

A) Flip Mutation

In flip mutation with binary encoding, the bits selected for creating new offsprings are inverted. If the bit is 1, it is converted into bit 0. Similarly, if the bit is 0, it is converted into bit 1. Figure 2.19 illustrated flip mutation with binary encoding.

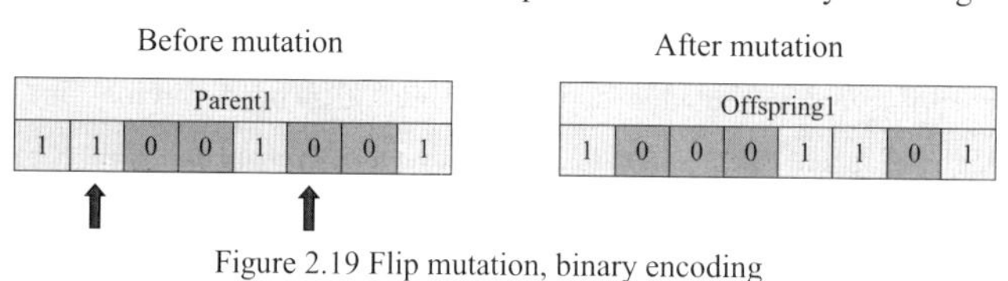

Figure 2.19 Flip mutation, binary encoding

III) Permutation encoding mutation

Here, we will propose swap mutation techniques with permutation encoding:

A) Swap Mutation

In swap mutation with permutation encoding, the order of the two numbers given in a sequence are exchanged. Figure 2.20 illustrated swap mutation with permutation encoding.

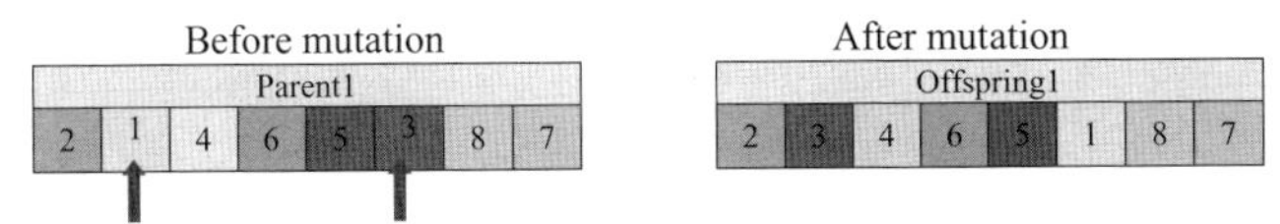

Figure 2.20 Swap mutation, permutation encoding

IV) Value encoding mutation

Here, we will propose swap mutation techniques with value encoding [46]:

A) Swap Mutation

In swap mutation with value encoding mutation, a small numerical value is either added or subtracted from the selected values of chromosomes to create new offsprings. Figure 2.20 illustrated swap mutation with value encoding mutation

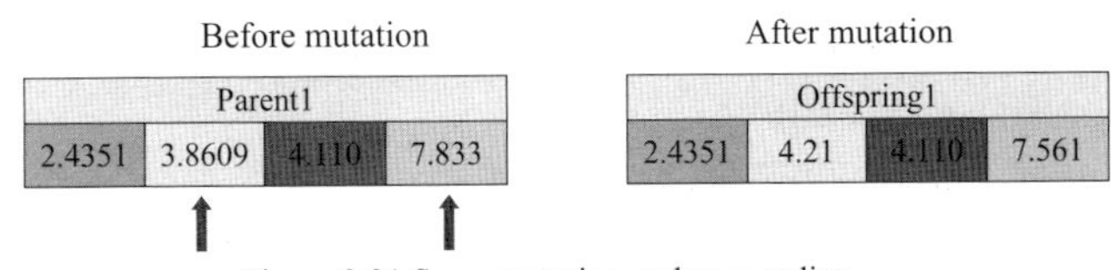

Figure 2.21 Swap mutation, value encoding

V) Tree encoding mutation

Here, we will propose swap mutation techniques with tree encoding [55]:

A) Swap Mutation

In swap mutation with tree encoding mutation, the certain selected nodes of the tree are muted to create new offspring, which is illustrated in Figure 2.21.

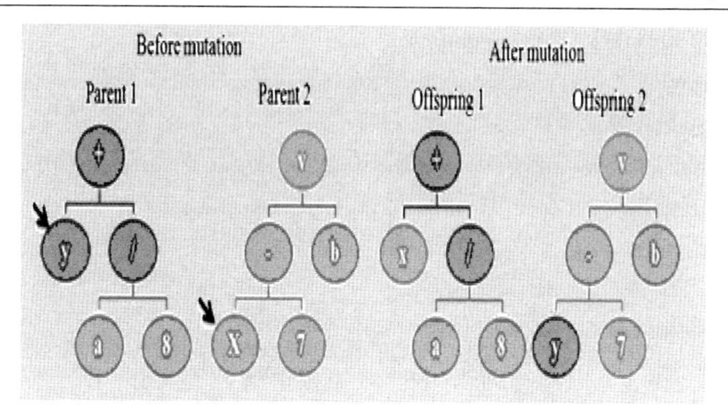

Figure 2.22 Swap mutation, tree encoding mutation

2.3.5 Repair

The main idea of the chromosome-repairing scheme is that, for an infeasible solution, repair it for transforming it into a feasible solution. For facilitating the illustration, the chromosome which represents an infeasible solution is called infeasible chromosome. Repairing a chromosome means a generating a feasible chromosome an infeasible one through some repairing procedure. Repairing strategy depends on the existence of a deterministic repair procedure. Three repair algorithms were implemented and they are described in the following [52]:

A) Lamarckian Approach

Lamarckian approach consist of genetically modify an infeasible solution and the transformed infeasible solution and the transformed infeasible solution, i.e., a feasible one, replaces the infeasible one in the population for further evolution.

B) Baldwinian Approach

A less destructive approach of the infeasible solutions allows combine learning and evolution. In this approach the solutions are repaired only for their evaluation. Analytic and empiric studies indicate that this technique reduces the speed of convergence of the evolutionary algorithm and allows converging to global optimum.

C) Annealing Approach

This approach is based on the main concepts involved in Simulated Annealing. The infeasible solutions are accepted with certain probability. At first stages the infeasible solutions are accepted then this probability of acceptation is decreasing. When a infeasible solution is not accepted it is repaired.

In all the above approaches, the process of repairing a solution consists in swapping the oil well that is located in an incorrect shift with a randomly selected oil well located in a different shift. This process is repeated to every oil well that not fulfill the problem constraint.

2.3.6 Migration

In this stage, we migrate new offspring with old population and create the new population taking the best ones of parents and children population using fitness function. Then we reserve the best chromosomes to the next generation.

2.3.7 Termination test

The algorithm is terminated when the maximum number of generations is achieved, or when the individuals of the population converges; Convergence occur when all individuals positions in the population are identical. In this case, crossover will have no further effect. Otherwise, create other new population

2.4 Genetic algorithm Parameters

The basic parameters of GA are crossover probability, mutation probability and population size [31].

2.4.1 Crossover Probability (Pc)

The crossover probability is defined as the ratio between the numbers of offspring produced by crossover operator in each generation to the population size. The crossover probability is denoted by Pc. If Pc equals 100%, then all offspring are produced by crossover. If it is 0%, then the completely new generation is produced from exact copies of chromosomes from old population.

Crossover is made in hope that new chromosomes will be better than old population.

The performance of GA depends, to a great extent, upon the crossover probability value. A crossover probability with higher probability allows exploration of more of the solution space, and reduces the chances of settling for a false optimum. But if crossover probability is too high, of a lot of computation time in exploring is wasted in unpromising regions of the solution space.

2.4.2 Mutation Probability (Pm)

Without mutation, offspring are directly copied from population after crossover without any change. If the mutation is performed, some chromosomes are changed. The mutation probability is defined as the ratio between numbers of offspring produced by mutation operator in each generation to the total number of genes in the population. The mutation probability is denoted by Pm. If Pm is 100%, all chromosomes are changed, if it is 0%, nothing is changed.

Generally, mutation is important to prevent the GA from falling into local extremes. The performance of GA depends also upon the mutation probability value. If Pm is too low, many useful genes are never tried out. If it is too high, there will be much random perturbation, the offspring will start to lose their resemblance to the parents, and the algorithm will not learn from the history of the search [46].

2.4.3 Population Size

One of basic parameters of GA is population size. Minimum numbers of chromosomes in one generation are required to reach as fast as to the optimal solution. If chromosomes are too few, GA will have few possibilities to perform crossover and only a small part of search space is explored. On the other hand, if chromosomes are too many. Research shows that after some limit. This limit depends mainly on encoding and the problem. Using very large populations is not useful and it does not solve the problem faster than moderate sized

2.5 Advantages and disadvantages of GA

GA represents an efficient global method for optimization problems that are encountered in the earth sciences. GA has large advantages over classical techniques as; it does not require linearization assumptions or calculation of partial derivatives. The additional advantage is that the sampling is global, rather than local. However this advantage, GA has some drawbacks. Sometimes, GA has trouble finding the exact global optimum and there is no guaranty to find best solution. Another GA may take long time to evaluate the individuals. Here, we introduce some of its advantages and disadvantages.

2.5.1 Advantages of GA

The major advantages of GA when applied to optimization problems are:

1. Adaptability

GA does not have much mathematical requirement regarding about the optimization problems. GA can handle any kind of objective functions and any kind of constraints, i.e., linear or nonlinear, differential or non-differential and discrete search spaces.

2. Robustness

The use of GA operators makes it very effective in performing as global search, while most of conventional techniques usually a local search. GA is more efficient and more robust in locating optimal solution and reduces computational effort than other conventional techniques.

3. Flexibility

GA provides us great flexibility to be hybridized with other optimization methods to make an efficient implementation for a specific problem.

In addition these, Genetic algorithms can be applied to domains in which insufficient knowledge of the system and/or high complexity is there. Genetic algorithms are very effective techniques and it can quickly find a reasonable solution to a complex problem.

2.5.2 Disadvantages of GA

The disadvantage of genetic algorithms is that it, Sometimes, have a trouble in finding the exact global optimum because there is no guaranty to find best solution. Another drawback that GAs require large number of response (fitness) function evaluations depending on the number of individuals and the number of generations. In order to achieve good results using them, several design choices need to be cautiously decided in advance. For example, the population size, the recombination and mutation probability and the maximum number of generations. Choosing the correct parameters for GA is far from being a straight forward process, and requires practical expertise, as the success of certain GA parameters and operators cannot be generalized. In addition, the large population of solutions that gives the GA its power is also its bane when it comes to speed on a serial computer—the cost function of each of those solutions must be evaluated.

CHAPTER 3

A Chaos-based Evolutionary Algorithm for General Nonlinear Programming Problems

3.1 Introduction

Nonlinear programming problems are very important and frequently appear in the real world applications, such as structural optimization, engineering design, very-large-scale cell layout design, economics, resource allocation and many other applications [3,53]. Traditionally, the NLP problem is divided into two large classes: unconstrained optimization problems and constrained optimization problems, involving at least one inequality or equality constraint [54]. Unfortunately, there is no known method of determining the global optimal to the general nonlinear programming problem. For the unconstrained optimization problems there are many techniques which are classified to direct search methods and gradient based methods, while the constrained optimization problems algorithms are classified to an indirect and direct methods. All of these methods are called classical optimization techniques, which are local in scope, depending on the existence of derivatives, and they are insufficiently robust in discontinuous, vast multimodal, and noisy search spaces [54].

Some optimization methods that are conceptually different from the classical techniques have been appeared labeled as advanced techniques and are emerging as popular methods for the solution of complex engineering problems. These methods are based on certain characteristics and behavior of biological, molecular, swarm of insects, and neurobiological systems. Furthermore, advanced optimization techniques overcome difficulties and limitations of classical techniques and are less susceptible to getting 'stuck' at local optimal. In addition they require fewer parameters without requiring the objective function to be derivable or even continuous [7].

Among the existing advanced techniques, well-known algorithms such as (SA) [32, 33], (GA) [55, 56], (PSO) [29, 57], (ACO) [58,59], neural-network-based methods [34,35], and fuzzy optimization [60,61] etc. GA is one of the advanced methods and is presented as an efficient global method for nonlinear programming problems. GAs are well suited for solving such problems and it enjoys an increasing interest in the optimization community and many industrial applications [62-64]. New researchers introduced improved methods based on the hybridizing algorithms with genetic algorithms to improve its results. For instance, Tsoulos [65] introduced a heuristic modified method based on the genetic algorithm for solving constrained optimization problems. Juan and Ping [66] optimized the fuzzy rule base with combination of the GA and ant colony. Their results proved that the hybrid method can be more useful than the basic GA. Additionally, Sun and Tian [67] developed an efficient hybrid method for image classification with PSO and GA; where the authors used features of fast convergence of PSO and diversity of GA to improve the results quality.

On the other hand, GAs can escape from local optima traps and find the global optima regions. However, their intensification process near the optimum set is often inaccurate. Thus, various approaches have been created to improve the local search capability of genetic algorithm by hybridizing it with local search techniques [68-73]. Donis-Díaz et al. [68] introduced a hybrid model of GA with local search and show that the hybrid model improves the results compared to those obtained by using the classical model of GA. Sawyerr and Adewumi [69] improved the local search capability of GA by hybridizing real coded genetic algorithm with 'uniform random' local search. In [70], Yang et al. developed a hybrid algorithm which is a combination between genetic algorithm and local search to solve the parametric mixed-integer programming problem; where GA is used to perform global search, while LS strategy is applied to each generated individual (or chromosome) of the population.

Furthermore, Derbel et al. [71] proposed a GA combined with an iterative local search to improve the solutions generated by GA and intensify the search space. In addition, Derbel et al. show by numerical experiments that hybrid algorithm improves the best known solutions previously obtained by the tabu search heuristic. Kabir et al. [72] incorporated a new local search operation that is devised and embedded in hybrid genetic algorithm to fine-tune the search in feature selection process. Finally, in [73] Kilani make a Comparison between the performance of the genetic and local search algorithms for solving the satisfiability problems.

In the recent years, the mathematics of chaos theory has been applied to many aspects of the optimization sciences. Chaos theory was initially described by Hénon [74] and was summarized by Lorenz [75]. Chaos is a common nonlinear phenomenon in nature, where it is fully reflects the complexity of the system that will be useful in optimization. Chaotic maps can easily be implemented and avoid entrapment in local optimal [76-80]. As a novel method of global optimization, chaos optimization algorithms have attracted much attention, which were all based on Logistic map. The inherent characteristics of chaos can enhance optimization algorithms by enabling it to escape from local solutions and increase the convergence to reach to the global solution. For instance, in [81] an experimental analysis on the convergence of evolutionary algorithms (EAs) is proposed; where the effect of introducing chaotic sequences instead of random ones during all the phases of the evolution process is investigated. This approach is based on the substitution of chaotic sequences with the random number generator. Tavazoei and Haeri in [82] proposed a new optimization technique by modifying a chaos optimization algorithm (COA) based on the fractal theory. Additionally, by considering the statistical property of the sequences of Logistic map and Kent map, Yang et al. [83] proposed an improved hybrid chaos-Broyden-Fletcher-Goldfarb-Shanno (BFGS) optimization algorithm and the Kent map based hybrid chaos-BFGS algorithm. While, in [84] Cong et al. proposed an improved fast convergent chaos optimization algorithm based on the ergodic and stochastic properties of the chaos variables; which is more effective in complex optimization problems.

Many researchers proposed integration between chaos theory and optimization algorithms to improve the solution quality. The authors in [85] presented hybrid chaos-PSO algorithm for the vehicle routing problem with time window. While, in [86] chaotic genetic algorithm based on Lorenz chaotic system for optimization problems was proposed. In [87] an improved quantum EA is presented based on PSO and chaos to avoid the disadvantage of easily getting into the local optional solution in the later evolution period. In [88] a new PSO methods that use chaotic maps for parameter adaptation is presented; where eight chaotic maps have been analyzed in the benchmark functions and twelve chaos-embedded PSO methods have been proposed. While, Zelinka et al. [89] discussed the mutual intersection of two interesting fields of research, i.e. evolutionary computation and deterministic chaos; where evolutionary computation are explored, and deterministic chaos is investigated as a behavioral part of EAs. Moreover, in [90] an effective self-adaptive differential evolution algorithm based on Gaussian probability distribution, gamma distribution and chaotic sequence (DEGC) for solving continuous global optimization problems is proposed. On the other hand, to increase the global search ability, chaotic sequences are applied in [91] to generate candidate solutions and a new searching mechanism is used to generate new solutions.

EAs are powerful computing tools to solve large-scale problems that have many local optima. However, they require high CPU times that are unpractical from the local optima. However, they require high CPU times that are unpractical from the engineering viewpoint, and they can escape from local optima traps and find the global optima regions. However, their intensification process near the optimum set is often inaccurate. On the other hand, local search schemes can converge quickly to these local minima and get stuck in a local optimum solution far away from the global optimal.

In this chapter, we introduce a brief view of chaos theory and some of its well-known maps. The main aim of this chapter is to propose our algorithm for solving NLP problem. The proposed algorithm is a chaos-based EA named chaotic genetic algorithm (CGA). The proposed algorithm is new optimization system that integrates genetic algorithm with a CLS strategy. The inherent characteristics of chaos can enhance optimization algorithms by enabling it to escape from local solutions and increase the convergence to reach to the global solution. Twelve chaotic maps have been analyzed in the proposed approach. By this way, it is intended to enhance the global convergence and to prevent sticking on a local solution. It has been detected that coupling with chaotic mapping save computational time and speed the convergence to the global solution. It has been also shown that, these methods especially logistic mapping have increased the solution quality that is in all cases they improved the global searching capability by escaping the local solutions. The simulation results of various numerical studies have been demonstrated the superiority of the proposed approach to finding the global optimal solution.

3.2 Chaos Theory

Chaos theory was initially described by Henon (1976) [74] and was summarized by Edward Lorenz (1993) [75]. It is a study in mathematics that has applications in several disciplines: meterorology, sociology, physics, engineering, economics, biology, and philosophy. Chaos theory refers to the study of chaotic dynamical systems as weather patterns, ecosystems, water flows, anatomical functions, or organizations. Chaotic systems are nonlinear dynamical systems that are highly sensible to their initial conditions. In other words, small changes in initial conditions result in high changes in the final outcome of system described as the so-called 'butterfly effect' detailed by Lorenz. One might think that chaos systems behave randomly, but a system does not necessarily need randomness for providing chaos behavior [92, 93]. In other words, deterministic systems are also able to show chaos behaviors. Recently, these characteristics have been utilized in optimization. There are three main properties of the chaotic behavior :

1) Periodicity
2) Randomness
3) Sensitivity to initial conditions

The periodicity property of chaos can ensure chaotic variables to traverse all non-repeated state within a certain range according to its own laws. So, chaos theory can be used as a global optimization mechanism. Chaos being radically different from statistical randomness, especially the inherent ability to search the space of interest efficiently, can improve the performance of optimization procedure. Chaos theory could be introduced into the optimization strategy to accelerate the optimum seeking operation and find the global optimal solution. The sensibility to the initial state is one of the most important characters of chaotic systems. This character ensures that there are no two identical new solutions obtained and maintains the population diversity.

3.3 Chaotic maps

Chaotic map is a map that exhibits some sort of chaotic behavior. Maps can be parameterized by a discrete-time or a continuous-time parameter. Discrete maps usually take the form of iterated functions. In this section, we offer some well-known chaotic maps found in the literature [92-100]

a) Chebyshev map

Chebyshev map is represented as [92]:

$$x_{t+1} = \cos\left(t\cos^{-1}(x_t)\right). \tag{3.1}$$

b) Circle map

Circle map is defined as the following representative equation [92]:

$$x_{t+1} = x_t + b - (a - 2\pi)\sin(2\pi x_t) \bmod(1); \tag{3.2}$$

where $a = 0.5$ and $b = 0.2$.

c) Gauss map

The Gauss map consists of two sequential parts defined as [93]:

$$x_{t+1} = \exp\left(-\alpha x_t^2\right) + \beta; \tag{3.3}$$

where α and β are real parameters.

d) Intermittency map

The intermittency map [94] is formed with two iterative equations and represented as:

$$\begin{cases} \varepsilon + x_t + c x_t^n, & \text{if } 0 < x_t \le p \\ \dfrac{x_t - p}{1 - p}, & \text{elseif } p < x_t < 1 \end{cases} \tag{3.4}$$

where $c = \dfrac{1 - \varepsilon - p}{p^2}$, $n = 2.0$, and ε is very close to zero.

e) Iterative map

The iterative chaotic map with infinite collapses [95] is defined with the following as:

$$x_{t+1} = \sin\left(\frac{a\pi}{x_t}\right); \tag{3.5}$$

where $a \in (0,1)$

f) Liebovitch map

The proposed chaotic map [92] can be defined as:

$$x_{t+1} = \begin{cases} \alpha x_t, & 0 < x_t \le p_1 \\ \dfrac{p_2 - x_t}{p_2 - p_1}, & p_1 < x_t \le p_2 \\ 1 - \beta(1 - x_t), & p_2 < x_t \le 1 \end{cases}; \tag{3.6a}$$

where

$$\alpha = \frac{p_2\left(1-\left(p_2 - p_1\right)\right)}{p_1} \text{ and } \beta = \frac{\left(\left(p_2 - 1\right) - p_1\left(p_2 - p_1\right)\right)}{p_2 - 1} \tag{3.6b}$$

g) Logistic map

Logistic map [96] demonstrates how complex behavior arises from a simple deterministic system without the need of any random sequence. It is based on a simple polynomial equation which describes the dynamics of biological population [97].

$$x_{t+1} = cx_t\left(1 - x_t\right); \tag{3.7}$$

where $x_0 \in (0,\ 1)$, $x_0 \notin \{0.0,\ 0.25,\ 0.50,\ 0.75,\ 1.0\}$ and when $c = 4.0$ a chaotic sequence is generated by the Logistic map.

h) Piecewise map

Piecewise map [97] can be formulated as follows:

$$x_{t+1} = \begin{cases} x_t / p, & 0 < x_t < p \\ \dfrac{x_t - p}{0.5 - p}, & p \le x_t < 0.5 \\ \dfrac{\left(1 - p - x_t\right)}{0.5 - p}, & 0.5 \le x_t < 1 - p \\ \dfrac{\left(1 - x_t\right)}{p}, & 1 - p < x_t < 1 \end{cases}; \tag{3.8}$$

where $p \in (0, 0.5)$ and $x \in (0, 1)$.

i) Sine map

Sine map [98] can be described as:

$$x_{t+1} = \frac{a}{4}\sin\left(\pi x_t\right); \tag{3.9}$$

where $0 < a \le 4$.

j) Singer map

One dimensional chaotic Singer map [99] is formulated as:

$$x_{t+1} = \mu\left(7.86x_t - 23.31x_t^2 + 28.75x_t^3 - 13.302875x_t^4\right); \tag{3.10}$$

where $\mu \in (0.9, 1.08)$.

k) Sinusoidal map

Sinusoidal map [100] is generated as the following equation:

$$x_{t+1} = ax_t^2 \sin(\pi x_t); \tag{3.11}$$

where $a = 2.3$.

l) Tent map

Tent map [91] is defined by the following iterative equation:

$$x_{t+1} = \begin{cases} x_t / 0.07, & x_t < 0.7 \\ \frac{10}{3}(1.0 - x_t), & x_t \geq 0.7. \end{cases} \tag{3.12}$$

3.4 The proposed algorithm

In this section we will discuss the proposed approach to solve a nonlinear programming problem, chaotic genetic algorithm (CGA), which is integration between GA, local search, and chaos theory. The proposed algorithm operates in two phases: in the first one, GA is implemented as global optimization system to find an approximate solution of the NLP problem. Then, in the second phase, CLS is introduced to accelerate the convergence and improve the solution quality. The basic idea of the proposed algorithm can be described as follows:

3.4.1 Phase I: GA

Step 1. Initial Population

Initially individuals are randomly generated to form an initial population, satisfying the entire range of possible solutions (the search space).

Step 2. Initial reference point

The algorithm needs at least one feasible reference point to enter the evolution mating, and gives offspring the best possible combination of the characteristics of process (i.e., complete the algorithm procedure), the interested reader is referred to Osman et al. [101].

Step 3. Repairing

This approach co-evolves the population of infeasible individuals until they become feasible. The idea of this technique is to separate any feasible individuals in a population from those that are infeasible by repairing infeasible individuals [66].

Step 4. Fitness Evaluation

The evaluation stage calculates a ranking metric of chromosomes fitness for each individual, which then determines their survival to the next generation optimization progresses through finding genes that provide higher fitness for the chromosomes in which it is found.

Step 5. Create a new population

We create a new population from the current generation by using the operators (ranking, selection, crossover and mutation).

A) Ranking: Ranks individuals according to their fitness value, and returns vector containing the corresponding individual fitness value, in order to establish later the probabilities of survival that are necessary for the selection process.

B) Selection: Stochastic universal sampling (SUS) [102] will be used; where the most important concern in a stochastic selection is to prevent loss of population diversity due to its stochastic aspect.

C) Crossover: Crossover is used to vary individuals from one generation to the next; where it combines two individuals (parents) to produce a new individuals (offspring) with probability (*Pc*). There are several techniques of crossover, one-point crossover, two-point crossover, cut and splice, uniform crossover and half uniform crossover, etc. [66,102]. Here we will use one-point crossover involving splitting two individuals and then combining one part of one with the other pair. This method performs recombination between pairs of individuals and returns the new individuals after mating, and gives offspring the best possible combination of the characteristics of their parents.

D) Mutation: Premature convergence is a critical problem in most optimization techniques, which occurs when highly fit parent individuals in the population breed many similar offspring in early evolution time. Mutation is used to maintain genetic diversity from one generation of a population to the next. In addition, mutation is an operator to change elements in a string which is generated after crossover operator [102, 103]. In this study, we will use real valued mutation; which means that randomly created values are added to the variables with a low probability (P_m).

Step 6. Migration

In this step, migrate of the new offspring with the old population to create the new population by taking the best individuals of parents and offspring population [104].

Step 7. Termination criteria

The algorithm is terminated when either the maximum number of generations has been produced, or the maximal number of evaluations is achieved.

3.4.2 Phase II: Chaotic local search

Optimization of the above-formulated objective function using GA yields an approximated optimal solution $x^* = (x_1^*, x_2^*, ..., x_n^*)$. Chaos-based local search has the ability to perturb x^*; where local region of x^* will be explored. The detailed description of CLS is described as follows:

Step 1. Determine variance range of chaotic search boundary

The range of CLS $[a_i, b_i]$, $i = 1, 2, .., N$ is determined by $x_i^* - \varepsilon < a_i$, $x_i^* + \varepsilon > b_i$; where ε is specified radius of chaos search.

Step 2. Generate chaotic variables

Perform chaos map by applying the chaos iteration operator. We select the Logistic Mapping method [105] which is used extensively using equation (3.7)

Step 3. Mapping chaos variable into the variance range

Chaos variable z^k is mapped into the variance range of optimization valuable $[a_i, b_i]$ by:

$$x_i^k = a_i + (b_i - a_i)z^k \tag{3.13}$$

which lead to

$$x_i^k = x_i^* - \varepsilon + 2\varepsilon z^k \quad \forall i = 1,...,n. \tag{3.14}$$

Step 4. Update the best value

If $f(x^k) < f(x^*)$ then set $x^* = x^k$, otherwise break the iteration.

Step 5. Stopping Chaos search

If $f(x^*)$ is not improved for all k iteration, stop chaos search process and put out x^* as the best solution. The pseudo code of the proposed CLS is shown in Figure 3.1, while Figure 3.2 shows the flow chart of the proposed algorithm.

```
CLS Procedure, given x* = (x1*, x2*, ....., xn*), ε and z^0.
    While: f(x*) is improved
        Begin
        k ← 1
            Generate z^k using different Chaotic maps
            x_i^k = x_i* − ε + 2εz^k    ∀i = 1,...,n
            If f(x^k) < f(x*) then x* = x^k
                Else if f(x^k) ≥ f(x*) continue,
            End if
            If termination criteria satisfied,
                Break
            End if
        k ← k + 1
    End while
```

Figure 3.1 Pseudo code of CLS

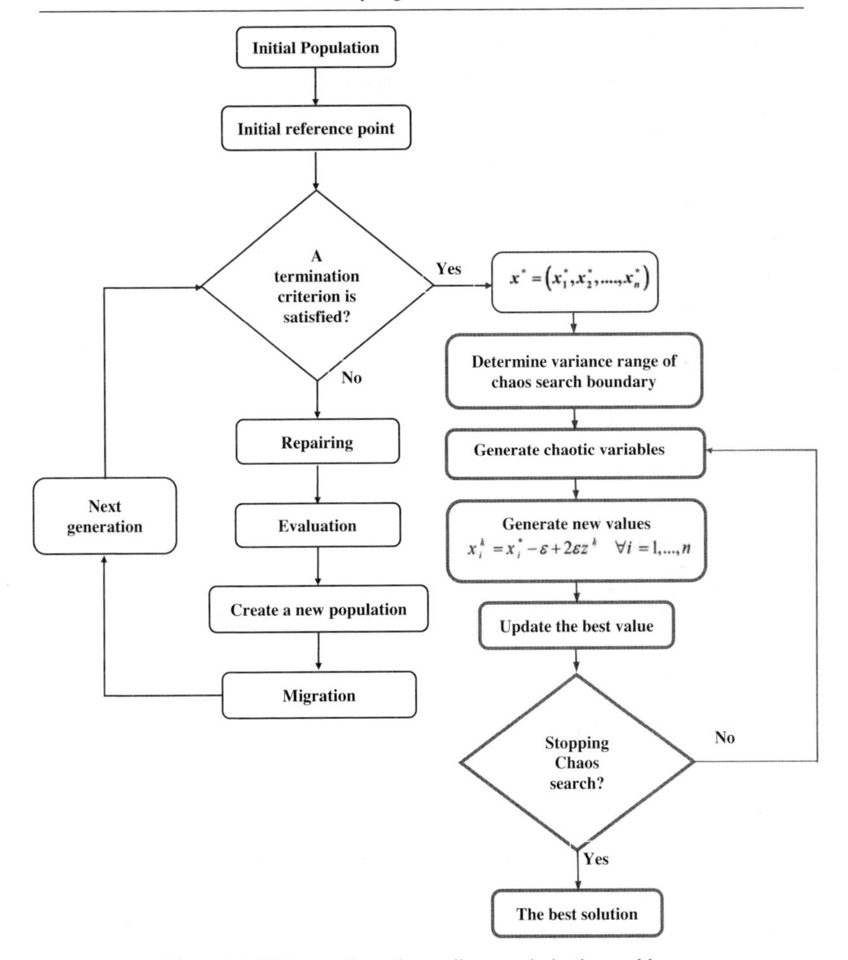

Figure 3.2 CGA procedures for nonlinear optimization problems

3.5 Experimental results

For evaluating the performance of the proposed algorithm for global optimization, CGA is tested on several well-known benchmark multimodal problems including the set of CEC'2005 special session on real parameter optimization and six constrained problems taken from the literature [106, 107]. The performance comparison with other optimization algorithms was done in order to demonstrate the efficiency and robustness of the proposed algorithm. CGA is coded in MATLAB 6.0 and the simulations have been executed on an Intel core (TM) i7-4500cpu 1.8GHZ 2.4 GHz processor. CGA, as any advanced algorithms, involves a number of parameters that affect the performance of algorithm. The parameters adopted in the implementation of CGA are listed in Table 3.1.

Table 3.1: CGA parameters

Generation gap	0.9
Crossover rate	0.9
Mutation rate	0.07
Selection operator	Stochastic universal sampling
Crossover operator	Single point
Mutation operator	Real-value
GA generation	50-1000
Chaos search iteration	1E02
Specified neighborhood radius	1E-6

3.5.1 Test function

3.5.1.1 Unconstrained benchmark problems

The proposed algorithm is tested by 25 unconstrained test problems [106] of dimension 10that appeared in the CEC'2005 special session on real parameter optimization. This suite is composed of the following functions [106].

- 5 unimodal functions

- F1: Shifted Sphere Function.
- F2: Shifted Schwefel's Problem 1.2.
- F3: Shifted Rotated High Conditioned Elliptic Function.
- F4: Shifted Schwefel's Problem 1.2 with Noise in Fitness.
- F5: Schwefel's Problem 2.6 with Global Optimum on Bounds.

- 20 multimodal functions

- 7 basic functions.

❖ F6: Shifted Rosenbrock's Function.

❖ F7: Shifted Rotated Griewank Function without Bounds.

❖ F8: Shifted Rotated Ackley's Function with Global Optimum on Bounds.

❖ F9: Shifted Rastrigin's Function.

❖ F10: Shifted Rotated Rastrigin's Function.

❖ F11: Shifted Rotated Weierstrass Function.

❖ F12: Schwefel's problem 2.13.

- Expanded functions.

❖ F13: Expanded Extended Griewank's plus Rosenbrock's Function (F8F2)

❖ F14: Shifted Rotated Expanded Scaffers F6.

❖ 11 hybrid functions. Each one (F15 to F25) has been defined through compositions of 10 out of the 14 previous functions (different in each case).

All functions are displaced in order to ensure that their optima can never be found in the center of the search space. In two functions, in addition, the optima cannot be found within the initialization range, and the domain of search is not limited (the optimum is out of the range of initialization).

3.5.1.2 Constrained benchmark problems

Furthermore, the proposed methodology is applied on six constrained benchmark problems. These constrained benchmark problems are taken from l iterator [107]. Table 3.2 lists the variable bounds, objective function and constraints for all these problems.

3.5.2 Performance Analysis Using Different Chaotic Maps

To evaluate the performance of each chaotic map, statistical results for different chaotic maps is presented to show the convergence process of the proposed algorithm. Due to space limitation, the proposed algorithm with different chaotic maps is implemented on 9 of the benchmark problems (7 of the 25 CEC'2005 benchmark problems and 2 constrained problems).

Table 3.2: Constrained benchmark problems

Problem	Variable bounds	Objective function [Minimize $f(x)$] and constraints $C(x)$
P1	$x_1 \in [-10,10]$ $x_2 \in [-10,10]$	$f_1(x) = x_1^2 + x_2^2$ $C_1(x) = x_1 - 3 = 0$ $C_2(x) = 2 - x_2 \leq 0$
P2	$x_1 \in [-10,10]$ $x_2 \in [-10,10]$	$f_1(x) = \frac{1}{4000}(x_1^2 + x_2^2)^2 - \cos(\frac{x_1}{\sqrt{1}})\cos(\frac{x_2}{\sqrt{2}}) + 1$ $C_1(x) = x_1 - 3 = 0$ $C_2(x) = 2 - x_2 \leq 0$
P3	$x_1 \in [0.1,10]$ $x_2 \in [0,10]$	$f_1(x) = \frac{-\sin(2\pi x_1)^3 \sin(2\pi x_2)}{x_1^3(x_1 + x_2)}$ $C_1(x) = x_1^2 - x_2 + 1 \leq 0$ $C2(x) = 1 - x_1 + (x_2 - 4)^2 \leq 0$
P4	$x_1 \in [13,100]$ $x_2 \in [0,100]$	$f_1(x) = (x_1 - 10)^3 + (x_2 - 20)^3$ $C_1(x) = -(x_1 - 5)^2 - (x_2 - 5)^2 + 100 \leq 0$ $C2(x) = (x_1 - 6) + (x_2 - 5)^2 - 82.81 \leq 0$
P5	$x_1 \in [-1,1]$ $x_2 \in [-1,1]$	$f_1(x) = x_1^2 + (x_2 - 1)^2$ $g_1(x) = x_2 - x_1^2$
P6	$x_1 \in [78,100]$ $x_2 \in [33,45]$ $x_3 \in [27,45]$ $x_4 \in [27,45]$ $x_5 \in [27,45]$	$f_1(x) = 5.3578547x_3^2 + 0.8356891x_1x_5 + 37.293239x_1 - 40792.141$ $C_1(x) = 85.334407 + 0.0056858x_2x_5 + 0.006262x_1x_4 - 0.0022053x_3x_5 - 92 \leq 0$ $C_2(x) = -85.334407 - 0.0056858x_2x_5 - 0.006262x_1x_4 + 0.0022053x_3x_5 \leq 0$ $C_3(x) = 80.51249 + 0.0071317x_2x_5 + 0.0029955x_1x_2 + 0.0021813x_3^2 - 110 \leq 0$ $C_4(x) = -80.51249 - 0.0071317x_2x_5 - 0.0029955x_1x_2 - 0.0021813x_3^2 + 90 \leq 0$ $C_5(x) = 9.300961 + 0.0047026x_3x_5 + 0 - 0012547x_1x_3 + 0.0019085x_3x_4 - 25 \leq 0$ $C_6(x) = -9.300961 - 0.0047026x_3x_5 - 0.0012547x_1x_3 - 0.0019085x_3x_4 + 20 \leq 0$

The algorithm has been run 10 times for each chaotic map; where in the proposed approach each run stops either when the maximal number of evaluations (10^5) is achieved, or when the error obtained (the distance between the best individual found and the global optimum in terms of fitness value) is less than 10^{-8}. Table 3.3 give the statistical results of the average error of the chosen problems using different chaotic maps, while Figure 3.3 shows a Comparison between the different chaotic maps according to the rank of the best result obtained by each map.

Table 3.3: Statistical results of the average error of nine of the benchmark problems using different chaotic maps

Function	Optimization method using different chaotic maps	Best	Worst	Mean	Standard deviation
F1	Logistic map	0	0	0	0
	Sinusoidal map	8.43E-09	8.41E-09	8.42E-09	9.44E-12
	Chebyshev map	9.76E-09	9.68E-09	9.72E-09	4.15E-11
	Singer map	1.35E-08	1.34E-08	1.34E-08	2.19E-11
	Tent map	8.23E-07	8.17E-07	8.2E-07	3.12E-09
	Sine map	2.34E-06	2.33E-06	2.34E-06	4.01E-09
	Circle map	2.35E-03	0.002323	0.002332	1.1E-05
	Piecewise map	4.47E-07	4.44E-07	4.45E-07	1.24E-09
	Gauss map	8.90E-01	0.883827	0.885911	0.002907
	Intermittency map	9.97E-03	0.009889	0.009918	3.86E-05
	Liebovitch map	9.95E-04	0.000986	0.000989	4.56E-06
	Iterative map	6.63E-05	6.61E-05	6.62E-05	1.21E-07
F5	Logistic map	1.63E-03	0.001624	0.001626	2.98E-06
	Sinusoidal map	1.68E-03	0.001677	0.001678	1.7E-06
	Chebyshev map	1.99E+01	19.73715	19.80133	0.082031
	Singer map	2.34E-01	0.23226	0.233024	0.000872
	Tent map	2.44E-02	0.02434	0.024369	2.99E-05
	Sine map	4.34E-03	0.004329	0.004334	5.55E-06
	Circle map	8.92E-02	0.089094	0.089079	8.49E-05
	Piecewise map	3.33E-03	0.003306	0.003318	1.22E-05
	Gauss map	1.64E-03	0.001632	0.001633	1.99E-06
	Intermittency map	1.98E-03	0.001979	0.001979	1.83E-06
	Liebovitch map	1.63E-03	0.001632	0.001632	1.22E-06
	Iterative map	9.99E-03	0.009979	0.009975	6.94E-06
F9	Logistic map	0.00E+00	0	0	0
	Sinusoidal map	9.55E-11	9.55E-11	9.55E-11	4.52E-14
	Chebyshev map	8.15E-13	8.1E-13	8.13E-13	2.38E-15
	Singer map	9.88E-08	9.84E-08	9.85E-08	1.67E-10
	Tent map	5.55E-06	5.54E-06	5.54E-06	2.89E-09
	Sine map	7.90E-10	7.86E-10	7.88E-10	1.67E-12
	Circle map	7.90E-05	7.89E-05	7.89E-05	6.18E-08
	Piecewise map	9.90E-07	9.84E-07	9.86E-07	2.94E-09
	Gauss map	9.65E-09	9.62E-09	9.63E-09	1.62E-11
	Intermittency map	4.57E-11	4.55E-11	4.56E-11	1.03E-13
	Liebovitch map	8.87E-07	8.84E-07	8.85E-07	1.22E-09
	Iterative map	7.29E-09	7.25E-09	7.27E-09	1.86E-11
F13	Logistic map	3.36E-01	0.336	0.336	1.86798E-08
	Sinusoidal map	4.45E-01	0.445	0.445	1.09851E-07
	Chebyshev map	2.34E+01	23.40002	23.40001	1.03623E-05
	Singer map	1.13E+00	1.130001	1.13	3.17493E-07
	Tent map	4.67E+01	46.70001	46.70001	6.11585E-06
	Sine map	7.88E-01	0.788	0.788	3.73633E-08
	Circle map	3.78E+00	3.780001	3.780001	6.34022E-07
	Piecewise map	5.74E+00	5.740003	5.740002	1.27476E-06
	Gauss map	3.65E-01	0.365	0.365	3.66671E-08
	Intermittency map	5.55E+01	55.50001	55.50001	3.17136E-06
	Liebovitch map	8.44E+01	84.40008	84.40005	4.02895E-05
	Iterative map	4.01E-01	0.401	0.401	1.2911E-07

Table 3.3 (continued)

Function	Optimization method using different chaotic maps	Best	Worst	Mean	Standard deviation
F17	Logistic map	1.10E+02	110.0001	110	4.30333E-05
	Sinusoidal map	1.58E+02	158.0001	158	4.13631E-05
	Chebyshev map	1.21E+02	121.0001	121.0001	5.17368E-05
	Singer map	3.15E+02	315.0002	315.0001	0.000108114
	Tent map	1.21E+02	121.0001	121.0001	5.37776E-05
	Sine map	8.90E+03	8900.002	8900.002	0.001188764
	Circle map	1.23E+02	123.0001	123.0001	5.12232E-05
	Piecewise map	1.10E+01	11.00001	11.00001	4.5125E-06
	Gauss map	2.35E+02	235.0002	235.0001	0.000116965
	Intermittency map	5.43E+02	543.0002	543.0001	0.00010017
	Liebovitch map	5.64E+03	5640.001	5640.001	0.000664854
	Iterative map	1.11E+02	111	111	5.70794E-06
F21	Logistic map	3.00E+02	300.0002	300.0001	0.000108968
	Sinusoidal map	3.34E+02	334.0001	334	3.27361E-05
	Chebyshev map	5.32E+02	532	532.0001	3.1927E-05
	Singer map	5.64E+02	564.0002	564.0001	8.68057E-05
	Tent map	3.00E+02	300.0001	300.0001	6.5869E-05
	Sine map	2.91E+02	291.0003	291.0001	0.000126516
	Circle map	3.32E+02	332.0002	332.0001	8.49544E-05
	Piecewise map	8.92E+02	892.0009	892.0005	0.000429725
	Gauss map	7.98E+02	798.0006	798.0003	0.00032249
	Intermittency map	9.99E+02	999.0004	999.0002	0.000204309
	Liebovitch map	3.33E+02	333.0002	333.0001	0.000100067
	Iterative map	3.10E+02	310.0001	310.0001	7.43045E-05
F25	Logistic map	4.31E+02	431.0004	431.0002	0.000176739
	Sinusoidal map	4.44E+02	444.0002	444.0001	9.28503E-05
	Chebyshev map	4.56E+02	456	456	2.24657E-05
	Singer map	4.77E+02	477.0003	477.0002	0.000155603
	Tent map	4.99E+02	499.0001	499	2.93865E-05
	Sine map	5.00E+02	500.0003	500.0002	0.000153067
	Circle map	6.20E+02	620.0001	620.0001	5.58983E-05
	Piecewise map	7.10E+02	710.0001	710.0001	6.86835E-05
	Gauss map	9.10E+02	910.0005	910.0003	0.000244859
	Intermittency map	4.54E+02	454.0002	454.0001	8.65909E-05
	Liebovitch map	3.56E+02	356.0003	356.0002	0.000143809
	Iterative map	7.54E+02	754.0003	754.0002	0.000143744
P1	Logistic map	1.30E+01	1.30E+01	12.99429	0.005378
	Sinusoidal map	1.325E+01	1.325E+01	13.24836	0.00164
	Chebyshev map	1.30E+01	1.30E+01	13.00907	0.001086
	Singer map	1.31E+01	1.31E+01	13.09734	0.002393
	Tent map	1.30E+01	1.30E+01	13.01352	0.005991
	Sine map	1.31E+01	1.30E+01	13.04822	0.00093
	Circle map	1.31E+01	1.30E+01	13.05157	0.00325
	Piecewise map	1.31E+01	1.31E+01	13.09289	0.006423
	Gauss map	1.30E+01	1.30E+01	12.99441	0.004957
	Intermittency map	1.30E+01	1.30E+01	12.99606	0.003229
	Liebovitch map	1.30E+01	1.30E+01	12.99642	0.003314
	Iterative map	1.32E+01	1.32E+01	13.20288	0.00661

Table 3.3 (continued)

Function	Optimization method using different chaotic maps	Best	Worst	Mean	Standard deviation
P6	Logistic map	-3.07E+04	-3.07E+04	-30665.8	0.000275
	Sinusoidal map	-3.07E+04	-3.07E+04	-30665.7	7.79E-05
	Chebyshev map	-3.07E+04	-3.07E+04	-30665.8	0.000105
	Singer map	-3.07E+04	-3.07E+04	-30665.2	0.000883
	Tent map	-3.07E+04	-3.07E+04	-30665.5	0.001197
	Sine map	-3.07E+04	-3.07E+04	-30665.1	0.000506
	Circle map	-3.07E+04	-3.07E+04	-30665	0.000273
	Piecewise map	-3.07E+04	-3.07E+04	-30665.3	0.000296
	Gauss map	-3.07E+04	-3.07E+04	-30665.6	0.001168
	Intermittency map	-3.07E+04	-3.07E+04	-30665.6	0.001208
	Liebovitch map	-3.07E+04	-3.07E+04	-30665.6	0.000413
	Iterative map	-3.07E+04	-3.07E+04	-30665	0.001434

From the simulation result in Table 3.3 and Figure 3.3 , it is shown that the proposed algorithm with Logistic map can increase the quality of the solutions, avoid being trapped in local optima. In addition, we can see that the best performance is shown by the Logistic map; where it gave the best results.

3.5.3 Performance Analysis using logistic map

It has been also shown that, these methods especially logistic mapping have increased the solution quality that is in all cases they improved the global searching capability by escaping the local solutions. So, the Logistic map is applied in our algorithm. Figure 3.4 gives the sequence of 300 chaotic random numbers z^k generated in the Logistic map (Equation 10); where $z^0 = 10^{-3}$ by using the following equation.

$$z^k = \mu z^{k-1}\left(1 - z^{k-1}\right),\ z^0 \in (0,1),\ z^0 \notin \{0.0,\ 0.25,\ 0.50,\ 0.75,\ 1.0\},\ k = 1,2,... \quad (3.15)$$

For the 25 CEC'2005 benchmark problems, Table 3.4 illustrates comparison between the solutions obtained by CGA (after Phase I and after Phase II with Logistic map) and the global solution. From the Table, we can see that the solutions obtained by the proposed algorithm after CLS is better and converge to the optimal value than solutions obtained by the proposed algorithm before CLS.

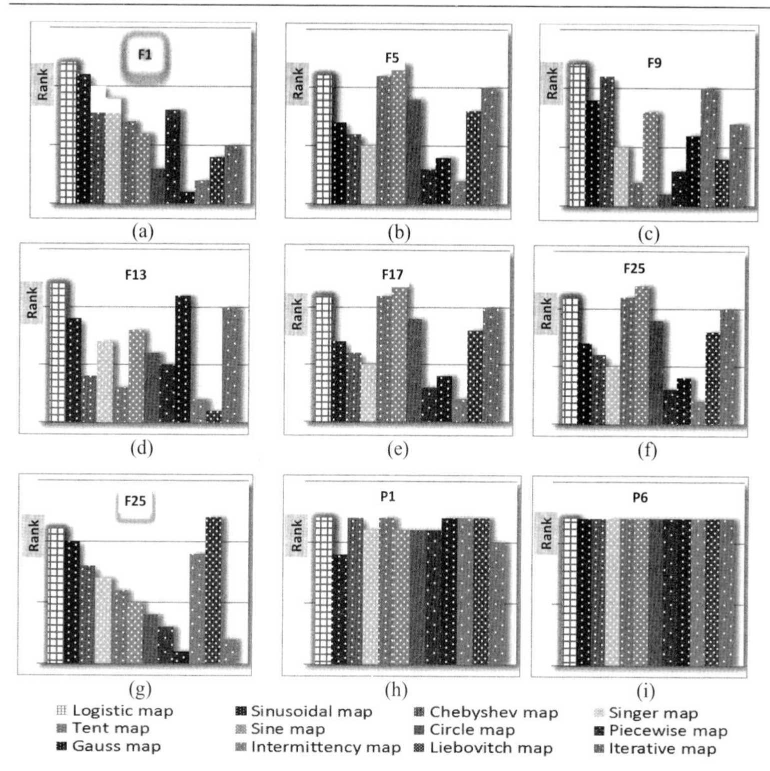

Figure 3.3 Comparison between the different chaotic maps according to the rank of the best result obtained by each map

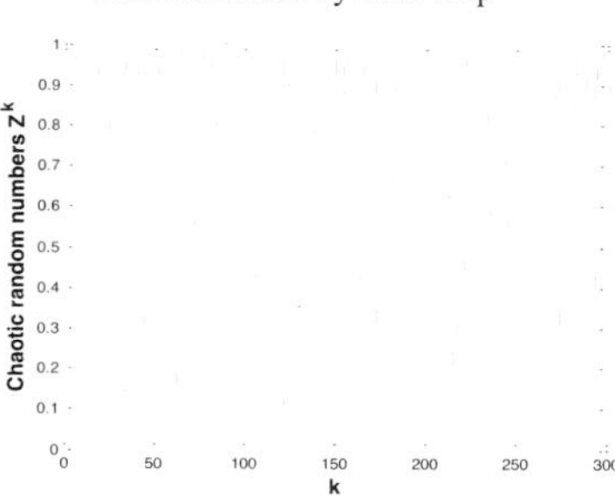

Figure 3.4 the chaotic random numbers

Table 3.4 Comparison between the global solution and CGA solution before and after CLS

Problem	Global solution	The proposed algorithm after Phase I	The proposed algorithm after Phase II	Problem	Global solution	The proposed algorithm after Phase I	The proposed algorithm after Phase II
1	-450.0000	-449.0942	-450.0000	**14**	-300.0000	-296.5665	-297.4822
2	-450.0000	-449.8730	-450.0000	**15**	120.0000	120.7922	120.0000
3	-450.0000	-426.4662	-427.3796	**16**	120.0000	241.9898	241.0303
4	-450.0000	-449.3676	-450.0000	**17**	120.0000	240.2219	239.6070
5	-310.0000	-309.9009	-309.9984	**18**	10.0000	448.0357	448.7040
6	390.0000	392.1299	391.1724	**19**	10.0000	310.8491	310.0000
7	-180.0000	-176.3826	-177.3475	**20**	10.0000	467.7710	467.6770
8	-140.0000	-118.9492	-119.9198	**21**	360.0000	561.6811	561.0725
9	-330.0000	-329.0428	-330.0000	**22**	360.0000	1.1132E+3	1.1124E+3
10	-330.0000	-323.5348	-324.0202	**23**	360.0000	920.2114	919.4683
11	90.0000	94.0769	93.9406	**24**	260.0000	460.3922	460.0000
12	-460.0000	-457.1980	-457.3399	**25**	260.0000	460.7327	460.2620
13	-130.0000	-129.2148	-129.6366				

Furthermore, the comparison between the average error obtained by CGA with Logistic map, and 9 continuous optimization algorithms [104-113] are reported in Table 3.5. All the algorithms have been run 50 times for each test function. From Table 3.5, we can say that CGA has produced better solutions than all 9 continuous optimization algorithms on average.

Table 3.5: Average error of the 25 CEC'05 benchmark functions obtained by our approach and 9 continuous optimization algorithms

Function	PSO [57]	IPOP-CMA-ES [58]	CHC [59,60]	SSGA [61,62]	SS-BLX [63]	SS-Arit [64]	DE-Bin [65]	DE-Exp [65]	SaDE [66]	CGA
F1	1.234E-4	0.000	2.464	8.420E-9	3.402E+1	1.064	7.716E-9	8.260E-9	8.416E-9	0.000
F2	2.595E-2	0.000	1.180E-2	8.719E-5	1.730	5.282	8.342E-9	8.181E-9	8.208 E-9	0.000
F3	5.174E+4	0.000	2.699E+5	7.948E+4	1.844E+5	2.535E+5	4.233E+1	9.935E+1	6.560E+3	22.6204
F4	2.488	2.932E+3	9.190E+1	2.585E-3	6.228	5.755	7.686 E-9	8.350E-9	8.087E-9	0.000
F5	4.095E+2	8.104E-10	2.641E+2	1.343E+2	2.185	1.443E+1	8.608E-9	8.514E-9	8.640 E-9	1.6E-3
F6	7.310E+2	0.000	1.416E+6	6.171	1.145E+2	4.945E+2	7.956 E-9	8.391 E-9	1.612E-2	1.724
F7	2.678E+1	1.267E+3	1.269E+3	1.271E+3	1.966E+3	1.908E+3	1.266E+3	1.265e+3	1.263E+3	2.653
F8	2.043E+1	2.001E+1	2.034E+1	2.037E+1	2.035E+1	2.036E+1	2.033E+1	2.038E+1	2.032E+1	2.008E+1
F9	1.438E+1	2.841E+1	5.886	7.286E-9	4.195	5.960	4.546	8.151E-9	8.330E-9	0.000
F10	1.404E+1	2.327E+1	7.123	1.712E+1	1.239E+1	2.179E+1	1.228E+1	1.118E+1	1.548E+1	5.9798
F11	5.590	1.343	1.599	3.255	2.929	2.858	2.434	2.067	6.796	3.9406
F12	6.362E+2	2.127E+2	7.062E+2	2.794E+2	1.506E+2	2.411E+2	1.061E+2	6.309E+1	5.634E+1	2.640
F13	1.503	1.134	8.297E+1	6.713E+1	3.245E+1	5.479E+1	1.573	6.403E+1	7.070E+1	0.363

Table 3.5 (continued)

Function	PSO [57]	IPOP-CMA-ES [58]	CHC [59,60]	SSGA [61,62]	SS-BLX [63]	SS-Arit [64]	DE-Bin [65]	DE-Exp [65]	SaDE [66]	CGA
F14	3.304	3.775	2.073	2.264	2.796	2.970	3.073	3.158	3.415	3.052
F15	3.398E+2	1.934E+2	2.751E+2	2.920E+2	1.136E+2	1.288E+2	3.722E+2	2.940E+2	8.423E+1	0.000
F16	1.333E+2	1.170E+2	9.729E+1	1.053E+2	1.041E+2	1.134E+2	1.117E+2	1.125E+2	1.227E+2	9.2745E+1
F17	1.497E+2	3.389E+2	1.045E+2	1.185E+2	1.183E+2	1.279E+2	1.421E+2	1.312E+2	1.387 E+2	1.196E+2
F18	8.512E+2	5.570E+2	8.799E+2	8.063E+2	7.668E+2	6.578E+2	5.097E+2	4.482E+2	5.320 E+2	4.387E+2
F19	8.497E+2	5.292E+2	8.798E+2	8.899E+2	7.555E+2	7.010E+2	5.012E+2	4.341E+2	5.195 E+2	3.000E+2
F20	8.509E+2	5.264E+2	8.960E+2	8.893E+2	7.463E+2	6.411E+2	4.928E+2	4.188E+2	4.767 E+2	4.576E+2
F21	9.138E+2	4.420E+2	8.158E+2	8.522E+2	4.851E+2	5.005E+2	5.240E+2	5.420E+2	5.140 E+2	2.010E+2
F22	8.071E+2	7.647E+2	7.742E+2	7.519E+2	6.828E+2	6.941E+2	7.715E+2	7.720E+2	7.655E+2	1.524E+2
F23	1.028E+3	8.539E+2	1.075E+3	1.004E+3	5.740E+2	5.828E+2	6.337E+2	5.824E+2	6.509E+2	5.595E+2
F24	4.120E+2	6.101E+2	2959 E+2	2.360E+2	2.513E+2	2.011E+2	2.060E+2	2.020E+2	2.000 E+2	2.000E+2
F25	5.099E+2	1.818E+3	1.764E+3	1.747E+3	1.794E+3	1.804E+3	1.744E+3	1.742E+3	1.738 E+3	1.00262E+2

In addition, we can rank the different algorithms according to the average error values and get Tables 3.6 and 3.7. Figure 3.5 illustrated the ranks obtained with CGA and 9 well-known algorithms [40,108-116]. Figures 3.6-3.30 show the convergence rate of the proposed approach for unconstrained problems. As result from Table 3.6, our approach has the first rank fifteen times and having the second rank three times. Finally, we can say that the proposed algorithm performs well, and more converges to the optimal solution of unconstrained test problems.

Table 3.6: Ranking of the average error values of the 25 CEC'05 benchmark functions for all algorithms.

Function	PSO [57]	IPOP-CMA-ES [58]	CHC [59,60]	SSGA [61,62]	SS-BLX [63]	SS-Arit [64]	DE-Bin [65]	DE-Exp [65]	SaDE [66]	CGA
F1	Rank 6	Rank 1	Rank 8	Rank 5	Rank 9	Rank 7	Rank 2	Rank 3	Rank 4	Rank 1
F2	Rank 6	Rank 1	Rank 9	Rank 5	Rank 7	Rank 8	Rank 4	Rank 2	Rank 3	Rank 1
F3	Rank 6	Rank 1	Rank 10	Rank 7	Rank 8	Rank 9	Rank 3	Rank 4	Rank 5	Rank 2
F4	Rank 6	Rank 10	Rank 9	Rank 5	Rank 8	Rank 7	Rank 2	Rank 4	Rank 3	Rank 1
F5	Rank 10	Rank 1	Rank 9	Rank 8	Rank 7	Rank 5	Rank 3	Rank 2	Rank 4	Rank 5
F6	Rank 8	Rank 1	Rank 9	Rank 5	Rank 6	Rank 7	Rank 2	Rank 3	Rank 4	Rank 5
F7	Rank 2	Rank 6	Rank 7	Rank 8	Rank 10	Rank 9	Rank 5	Rank 4	Rank 3	Rank 2
F8	Rank 10	Rank 1	Rank 5	Rank 8	Rank 6	Rank 7	Rank 4	Rank 9	Rank 3	Rank 2
F9	Rank 9	Rank 10	Rank 7	Rank 2	Rank 5	Rank 8	Rank 6	Rank 3	Rank 4	Rank 1
F10	Rank 6	Rank 10	Rank 1	Rank 8	Rank 5	Rank 9	Rank 4	Rank 3	Rank 7	Rank 1
F11	Rank 9	Rank 1	Rank 2	Rank 7	Rank 6	Rank 5	Rank 4	Rank 3	Rank 10	Rank 8
F12	Rank 9	Rank 6	Rank 10	Rank 8	Rank 5	Rank 7	Rank 4	Rank 3	Rank 2	Rank 1
F13	Rank 3	Rank 2	Rank 10	Rank 8	Rank 5	Rank 6	Rank 4	Rank 7	Rank 9	Rank 1
F14	Rank 8	Rank 10	Rank 1	Rank 2	Rank 3	Rank 4	Rank 6	Rank 7	Rank 9	Rank 5
F15	Rank 9	Rank 5	Rank 6	Rank 7	Rank 3	Rank 4	Rank 10	Rank 8	Rank 2	Rank 1

Table 3.6 (continued)

Function	PSO [57]	IPOP-CMA-ES [58]	CHC [59,60]	SSGA [61,62]	SS-BLX [63]	SS-Arit [64]	DE-Bin [65]	DE-Exp [65]	SaDE [66]	CGA
F16	Rank 10	Rank 7	Rank 1	Rank 3	Rank 2	Rank 6	Rank 4	Rank 5	Rank 8	Rank 9
F17	Rank 9	Rank 10	Rank 1	Rank 3	Rank 2	Rank 5	Rank 8	Rank 6	Rank 7	Rank 4
F18	Rank 9	Rank 5	Rank 10	Rank 8	Rank 7	Rank 6	Rank 3	Rank 2	Rank4	Rank 1
F19	Rank 8	Rank 5	Rank 9	Rank 10	Rank 7	Rank 6	Rank 3	Rank 2	Rank 4	Rank 1
F20	Rank 8	Rank 5	Rank 10	Rank 9	Rank 7	Rank 6	Rank 4	Rank 2	Rank 3	Rank 1
F21	Rank 10	Rank 2	Rank 8	Rank 9	Rank 3	Rank 4	Rank 6	Rank 7	Rank 5	Rank 1
F22	Rank 10	Rank 5	Rank 9	Rank 3	Rank 1	Rank 2	Rank 7	Rank 8	Rank 6	Rank4
F23	Rank 9	Rank 7	Rank 10	Rank 8	Rank 2	Rank 4	Rank 5	Rank 3	Rank 6	Rank 1
F24	Rank 8	Rank 9	Rank 7	Rank 5	Rank 6	Rank 2	Rank 4	Rank 3	Rank 1	Rank 1
F25	Rank 2	Rank 10	Rank 7	Rank 6	Rank 8	Rank 9	Rank 5	Rank 4	Rank 3	Rank 1

Table 3.7: Statistics of ranking.

Method	Rank1	Rank2	Rank3	Rank4	Rank5	Rank6	Rank7	Rank8	Rank9	Rank 10
CGA	15	3	0	2	3	0	0	1	1	0
PSO [57]	1	1	1	0	0	5	0	5	7	5
IPOP-CMA-ES [58]	7	2	0	0	5	2	2	0	1	6
CHC [59,60]	3	2	0	0	1	1	4	2	6	6
SSGA [61, 62]	0	2	2	1	5	1	3	8	2	1
SS-BLX [63]	1	2	4	0	4	5	4	3	1	1
SS-Arit [64]	0	2	0	4	1	7	5	2	4	0
DE-Bin [65]	0	3	4	9	3	3	1	1	0	1
DE-Exp [65]	0	5	8	4	1	1	3	2	1	0
SaDE [66]	1	2	6	6	2	2	2	1	2	1

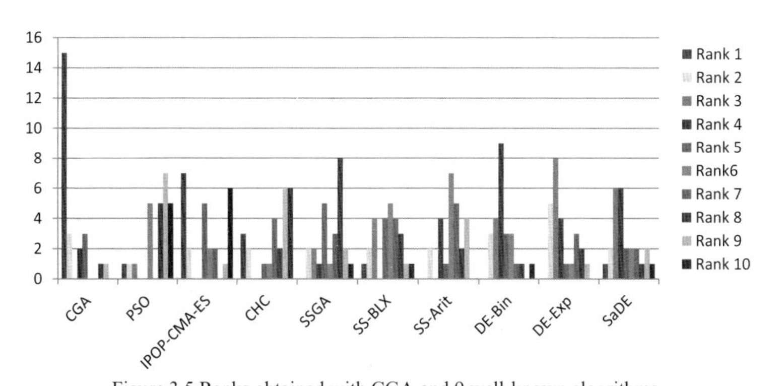

Figure 3.5 Ranks obtained with CGA and 9 well-known algorithms

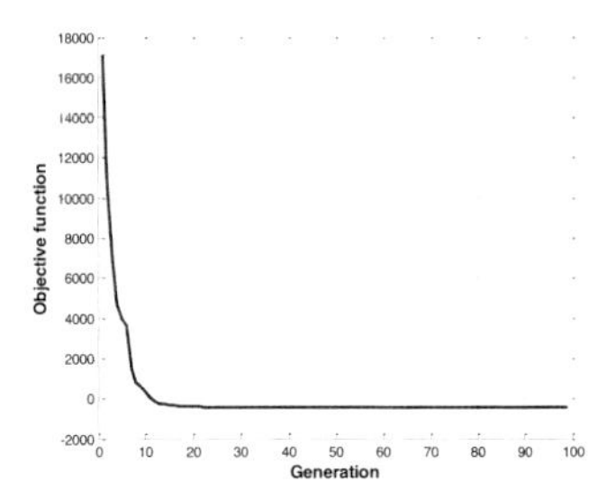

Figure 3.6 The convergence rate for F1

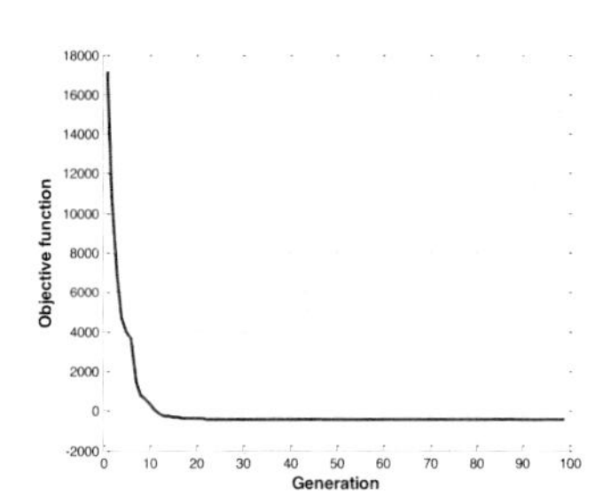

Figure 3.7 The convergence rate for F2

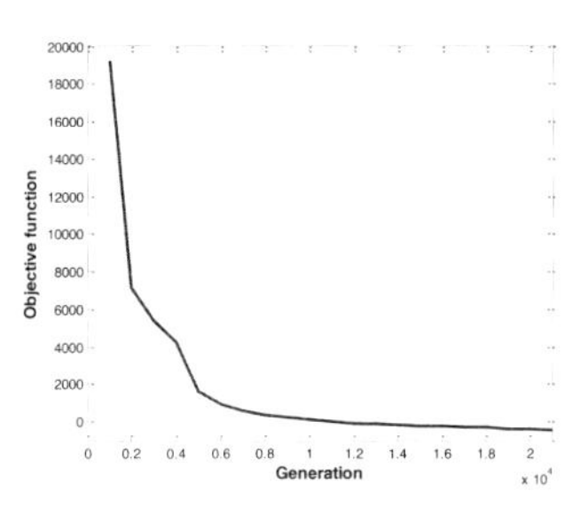

Figure 3.8 The convergence rate for F3

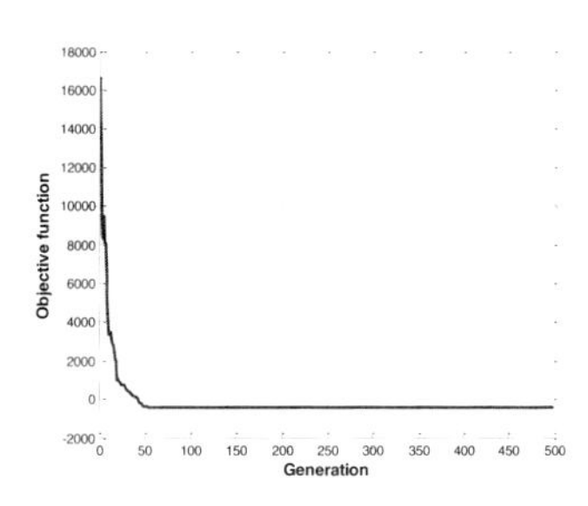

Figure 3.9 The convergence rate for F4

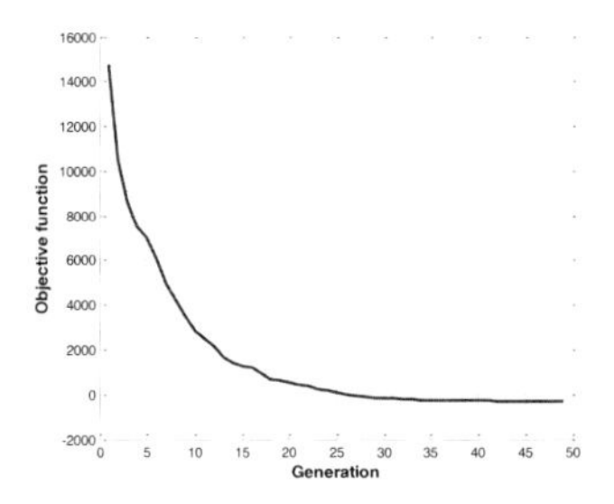

Figure 3.10 The convergence rate for F5

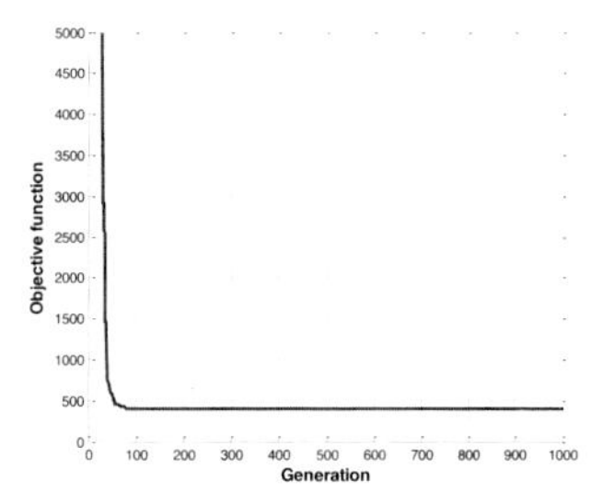

Figure 3.11 The convergence rate for F6

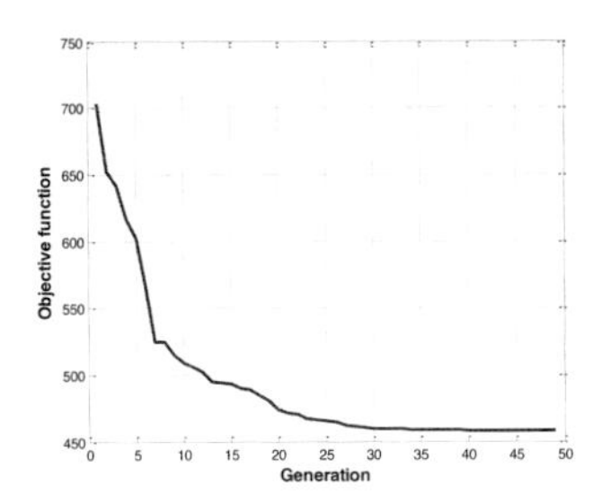

Figure 3.12 The convergence rate for F7

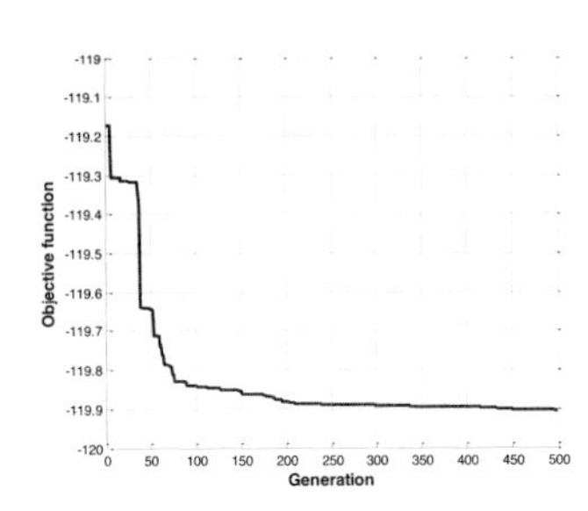

Figure 3.13 The convergence rate for F8

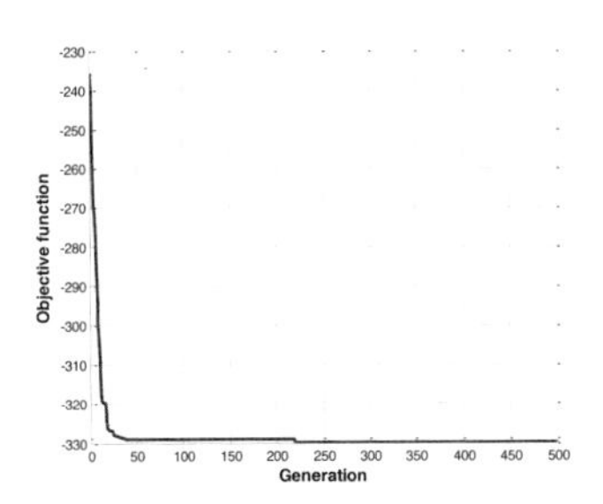

Figure 3.14 The convergence rate for F9

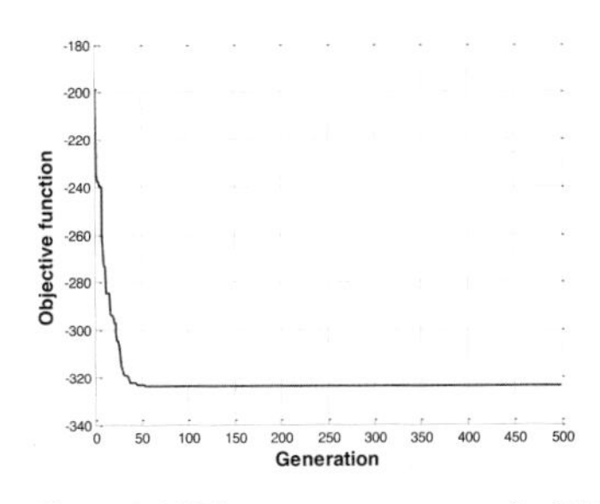

Figure 3.15 The convergence rate for F10

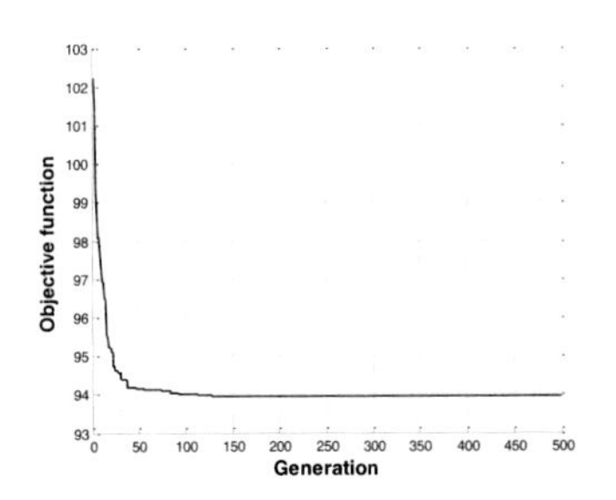

Figure 3.16 The convergence rate for F11

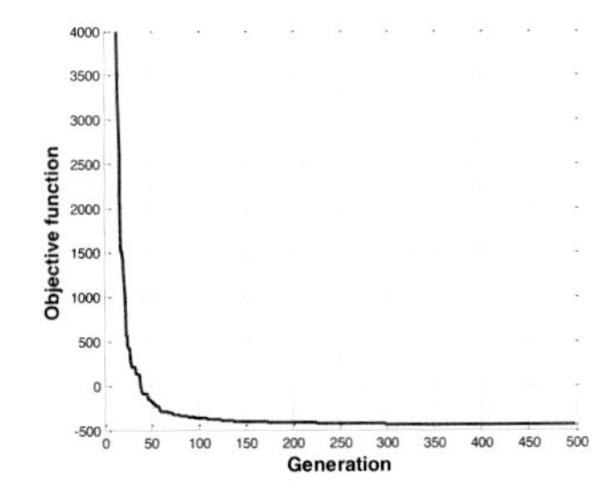

Figure 3.17 The convergence rate for F12

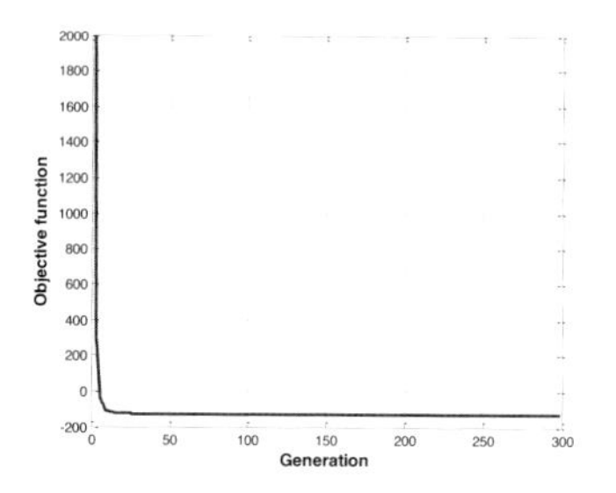

Figure 3.18 The convergence rate for F13

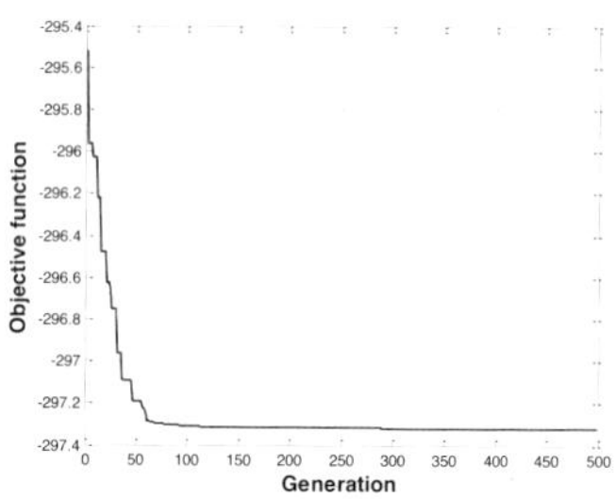

Figure 3.19 The convergence rate for F14

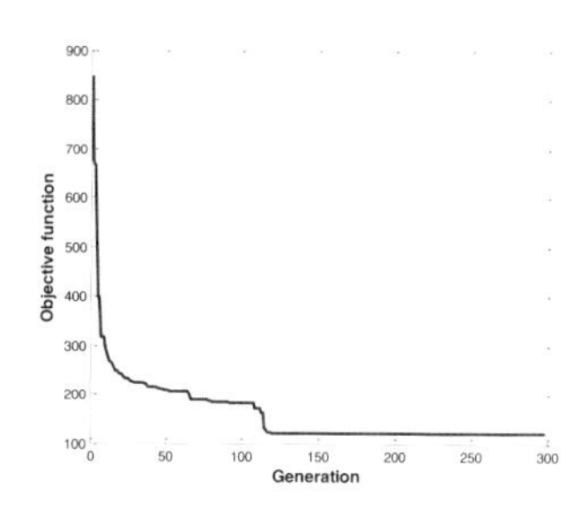

Figure 3.20 The convergence rate for F15

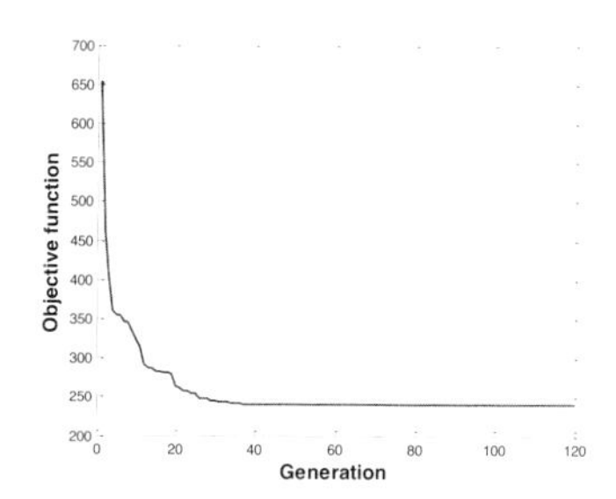

Figure 3.21 The convergence rate for F16

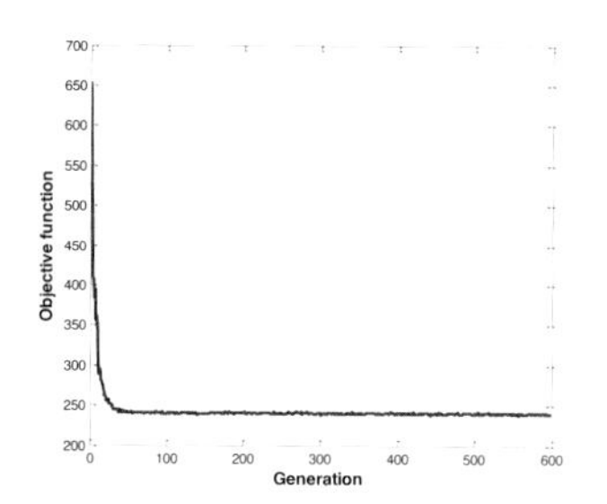

Figure 3.22 The convergence rate for F17

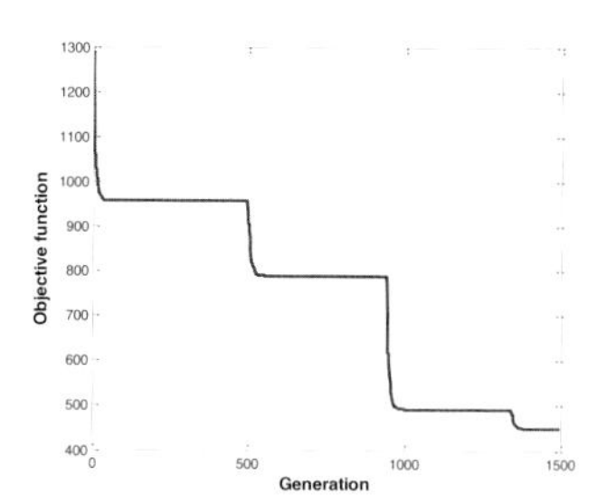

Figure 3.23 The convergence rate for F18

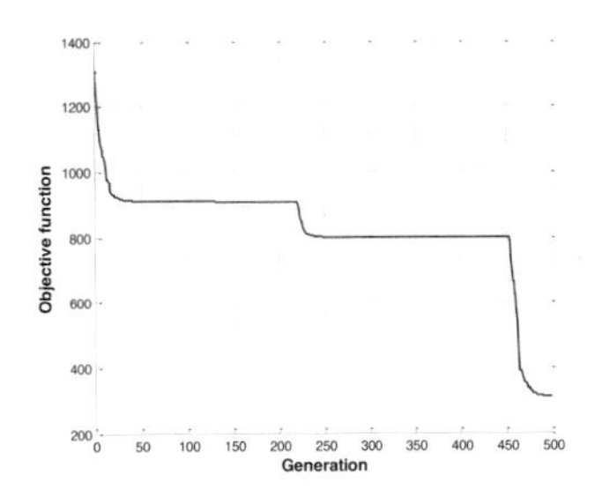

Figure 3.24 The convergence rate for F19

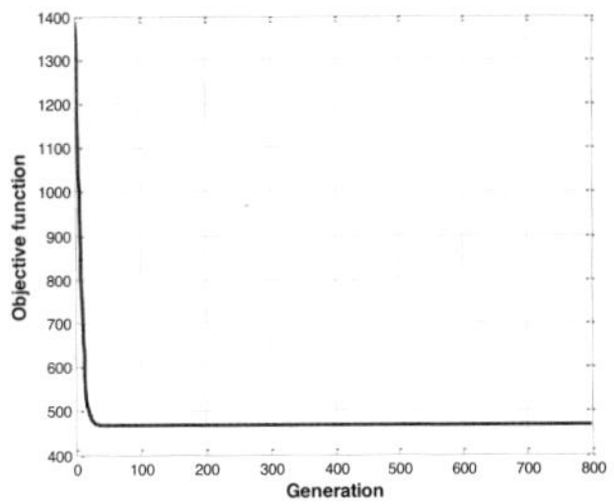

Figure 3.25 The convergence rate for F20

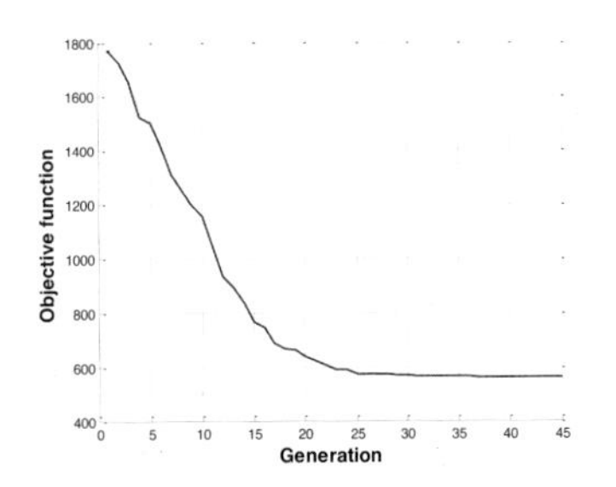

Figure 3.26 The convergence rate for F21

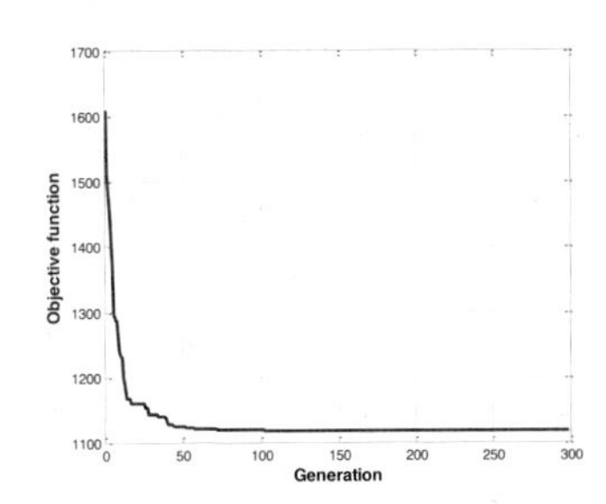

Figure 3.27 The convergence rate for F22

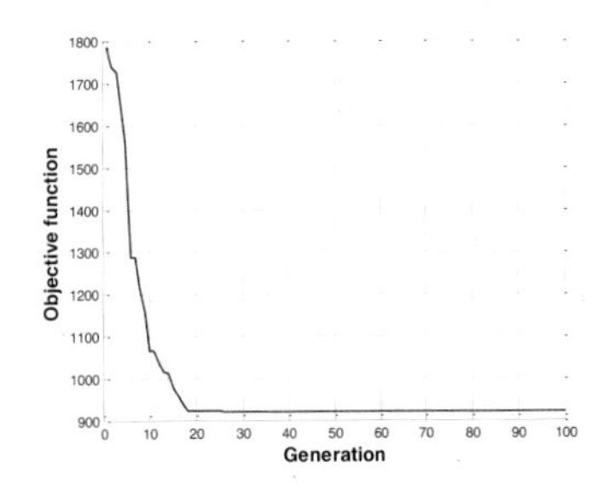

Figure 3.28 The convergence rate for F23

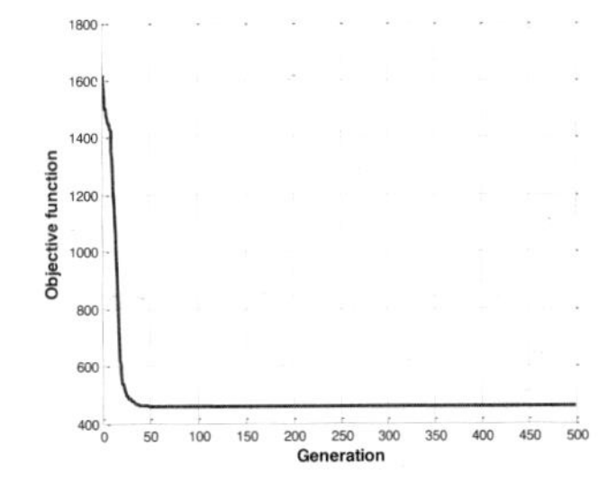

Figure 3.29 The convergence rate for F24

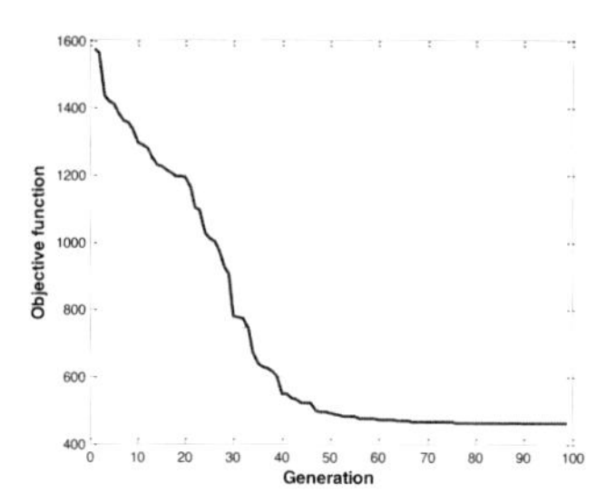

Figure 3.30 The convergence rate for F25

For the constrained problems, CGA and Augmented Lagrange Particle Swarm Optimization (ALPSO) [107] are applied to the 6 problems. In [107] Sedlaczek and Eberhard applied their method making 30 independent runs. Table 3.8 illustrates the comparison between the optimal solution, CGA result, and the best value obtained by ALPSO. As a result from Table 3.8, for the problems (P1, P2, P3 and P5), CGA is found the optimal solution and near to the optimal solution for the problem P4, and smaller than the optimal solution for the problem P6. On the other hand, ALPSO is found the optimal solution for the problems (P3, P5 and P6), near to the optimal solution for the problems (P1 and P2) and smaller than the optimal solution for the problem P3. So, CGA more converges to the optimal solution and exhibits a superior performance in comparison to ALPSO (i.e. CGA found the better solutions than ALPSO. Figures 3.31-3.36 show the convergence rate of the proposed approach for constrained problems

Table 3.8 The comparison between the optimal solution, the best value of ALPSO, and CGA result

Problem	Optimal solution	Best value of ALPSO [107]	CGA result before CLS	CGA result after CLS
P1	13.0000	12.9995	**14.0938**	**13.0000**
P2	0.01721	0.01719	**0.5742**	**0.01721**
P3	-0.09583	-0.09583	**1.8192**	**-0.09583**
P4	-6961.81	-6963.57	**-6959.9**	**-6961.804**
P5	0.75000	0.75000	**1.0652**	**0.750000**
P6	-30665.5	-30665.5	**-30664**	**-30665.8**

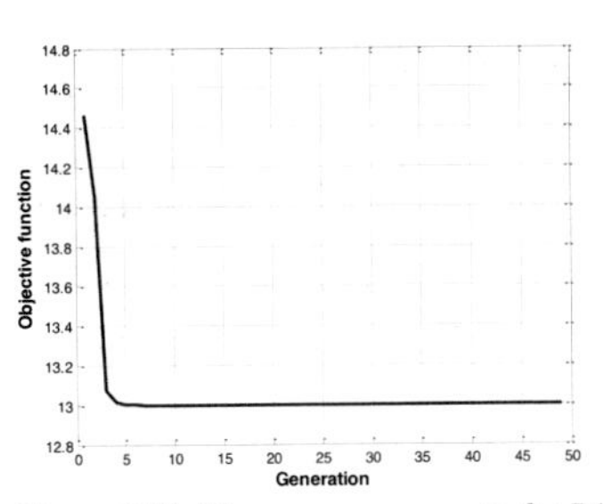

Figure 3.31 The convergence rate for P1

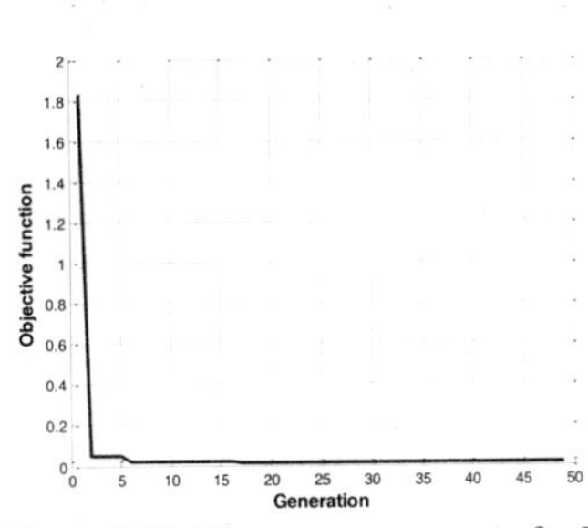

Figure 3.32 The convergence rate for P2

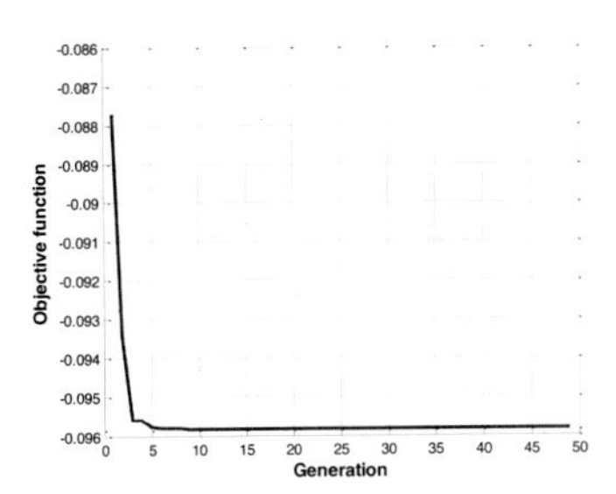

Figure 3.33 The convergence rate for P3

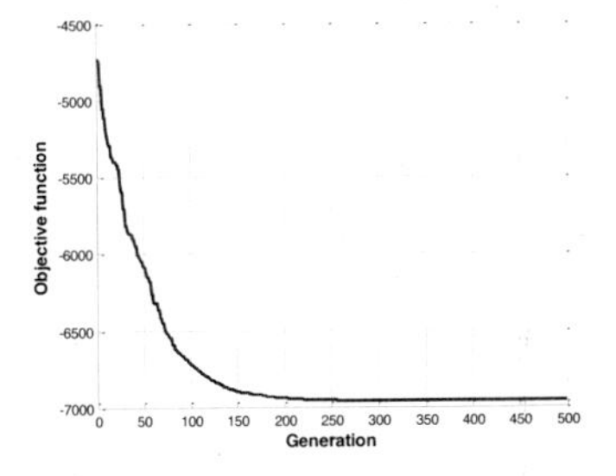

Figure 3.34 The convergence rate for P4

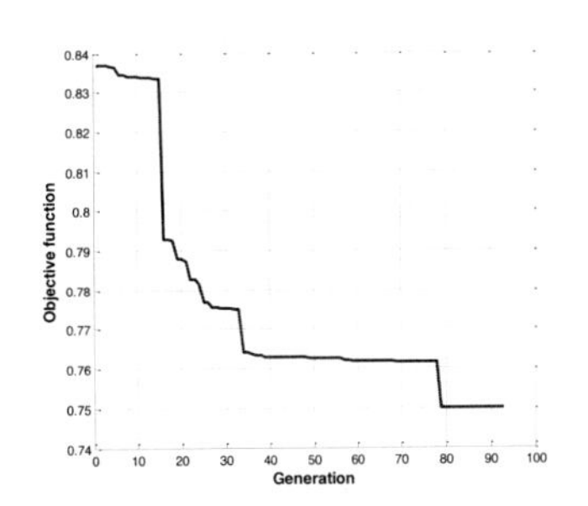

figure 3.35 The convergence rate for P5

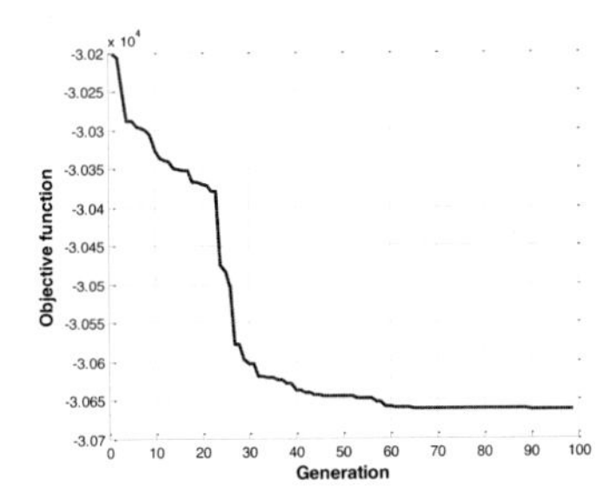

Figure 3.36 The convergence rate for P6

3.5.4 Speed Convergence analysis

In order to analyze the speed convergence behavior of the proposed algorithm and to provide computational time comparison, this section is presented. In order to compare the speed of convergence of the algorithm, a suitable measurement of computational time in needed. The number of iterations cannot be used as a reference of computation time as the algorithms perform different amount of works in their inner loops and have different population's sizes, here the number of fitness evaluations (FEs) was used as a measure of the computational time. In other words, the computational complexity of the algorithm was measured in order to better reflect its real running time. Table 3.9 shows the FEs consumed by CGA and the proposed approach without CLS in CEC 2005 test set. It has been also shown from the values reported in Table 3.9 that, CGA gives a significant reduction in computational time compared to the proposed approach without CLS.

Table 3.9 FEs consumed by CGA and the proposed approach without CLS in CEC 2005 test set

Function	FEs consumed by the CGA	FEs consumed by the proposed approach without chaotic local search	Saved (%)
F1	9.64E04	1.09E+05	11.6
F2	9.88E04	1.09E+05	9.36
F3	1E05	1.18E+05	15.3
F4	9.93E04	1.06E+05	6.32
F5	1E05	1.08E+05	7.41
F6	1E05	1.10E+05	9.09
F7	1E05	1.23E+05	18.7
F8	1E05	1.13E+05	11.5
F9	9.97E04	1.16E+05	14.1
F10	1E05	1.11E+05	9.91
F11	1E05	1.18E+05	15.3
F12	1E05	1.05E+05	4.76
F13	1E05	1.06E+05	5.66
F14	1E05	1.22E+05	18.0
F15	9.99E04	1.17E+05	14.6
F16	1E05	1.05E+05	4.76
F17	1E05	1.04E+05	3.85
F18	1E05	1.05E+05	4.76
F19	1E05	1.06E+05	5.66
F20	1E05	1.09E+05	8.26

Table 3.9 (continued)

Function	FEs consumed by the CGA	FEs consumed by the proposed approach without chaotic local search	Saved (%)
F21	1E05	1.10E+05	9.09
F22	1E05	1.02E+05	1.96
F23	1E05	1.10E+05	9.09
F24	1E05	1.00E+05	0.00
F25	1E05	1.17E+05	14.5

In addition, Figure 3.37 shows the Computation time saving percentage for CEC 2005 function set. It is shown from Figure 3.37 that our approach gives a significant reduction in computational time (9.34%) compared to the approach without CLS.

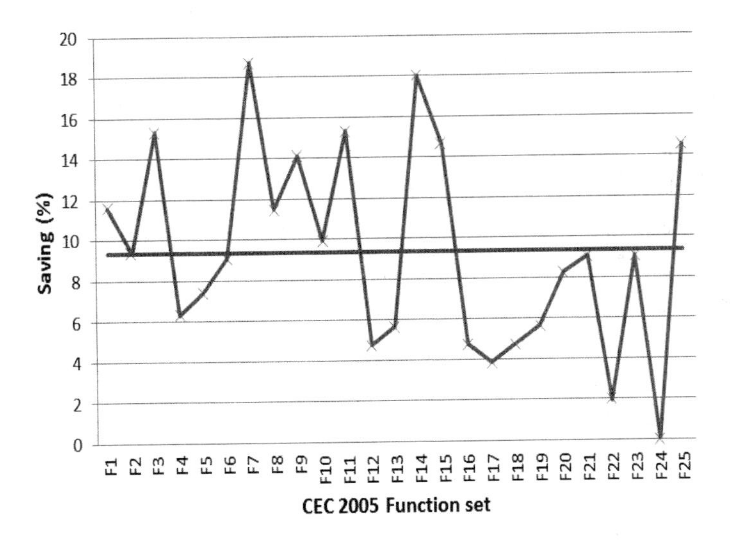

Figure 3.37 Computation time saving percentage for CEC 2005 function set

Finally, the main differences of some algorithms were presented in Table 3.10. As can be seen in Table 3.10, our proposed algorithm uses the advantages of chaotic maps and converge ability of EAs for solving optimization problems.

Table 3.10 Comparing some optimization algorithm

Features	Gradient based methods	Local search based methods	EAs	Chaotic methods	Proposed algorithm
High computational time	×	×	✓	×	×
Sensitive to initial conditions	✓	✓	✓	×	×
Need initial parameters	×	×	✓	×	✓
Global search capability	×	×	✓	✓	✓
Local search capability	✓	✓	×	×	✓

Computational complexity deals with determining the amount of effort needed to solve a certain problem. Evaluating computational complexity of the proposed algorithm involves searching a very large space; where the computational complexity for these problems is NP. The proposed approach can be used to reach an acceptable solution with acceptable running time in comparison with other methods. It is important to note that the proposed algorithm does not only guarantee achieving the best results but in a much shorter time than previous methods. The main reason of lower computational complexity of running the proposed algorithm in comparison with the other method, is that our approach avoiding the systematic search in the problem space. With a suitable choice of operators of genetic algorithm and chaotic maps, The proposed approach can be used to reach an optimal computational complexity with respect to the performance of genetic algorithm, certainly less than the computational complexity of other methods and can be around order of $O(n^2)$.

3.6 Conclusion

In this chapter we presented a Chaos-based genetic algorithm for solving nonlinear programming problems CGA. The proposed approach integrates genetic algorithm with a chaotic local search to accelerate the optimum seeking operation and find the global optimal solution. Implementing a local search referred to Chaos search incorporated with genetic algorithm, It can provide a more efficient behavior and a higher flexibility for nonlinear optimization problems. Technique is tested by solving constrained and unconstrained test problems and the resulted outcomes are analyzed. The research found that the results by hybridizing genetic optimization with chaos theory achieve better results. CGA introduces near optimal solution for unconstrained and constrained nonlinear optimization problems. CGA was demonstrated to be extremely effective and efficient at locating optimal solutions.

CHAPTER 4

Job Shop Scheduling Problems

4.1 Introduction

Production scheduling is an essential and intangible factor of the logistical performance of production organizations. It becomes a critical factor in many job shops in order to determine their capacity for more work and be able to schedule their work more efficiently. Big number of research has been carried out in the field of production scheduling [117]. Nevertheless, scheduling tasks can become very complex for humans. Practitioners in production scheduling often are convinced that scheduling leaves much room for improvement. In practice, formal techniques are rarely used straightforwardly, and schedulers mostly still use their own "rules–of–thumb," especially in dynamic, uncertain and complex scheduling environments. JSSP is a branch of production scheduling, which is one of the hardest combinatorial optimization problem. JSSP occurrence is not confined to single production shops. But also, it can be used to model a large set of structurally equivalent or similar problems manufacturing planning, in supply web logistics, resource bounded project planning, task assignment and so forth [117,118].

In the classical JSSP, n jobs are processed to completion on m unrelated machines. Each job requires processing on each machine exactly once. For each job, technology constraints specify a complete, distinct routing, which is fixed and known in advance. Processing times are sequence-independent, fixed, and known in advance. Each machine is continuously available from time zero, and operations are processed without interruption (non-preemption). The common objective is to minimize the maximum completion time (makespan). Each machine is limited to executing one operation (one step of the job route) at a time. Each job has a release-date (the time after which the operations in the job may be executed) and a due-date (the time by which the last activity in the job must finish). A schedule is an assignment of a start time to each operation to find the shortest possible schedule that satisfies the constraints of all the precedence, machines, release-date, and due-date [190, 120]. Since each job is partitioned into the operations processed on each machine, a schedule for a certain problem instance consists of an operation sequence for each machine involved. These operation sequences can be permuted independently from each other. There are a total of $(n!)^m$ different sequences. Therefore, scheduling problems are NP; NP stands for non-deterministic polynomial. NP means that it is not possible to solve an arbitrary instance in polynomial time [120,121].

A large number of approaches have been reported in the Operations Research (OR) literature to model and solve scheduling problems. These approaches achieve varied degrees of success and are turned around a series of technological advances that have occurred over the last years. Qing and Wang [122] propose a new hybrid genetic algorithm for JSSP. Vilcaut and Billaut [123] introduce a tabu search and a genetic algorithm for solving a bicriteria general JSSP. Ghiani, Grieco, et al [124] propose a fast heuristic for large-scale assembly JSSP with bill of materials. Gromicho, Hoorn and et al [125] solve the JSSP optimally by dynamic programming. Hasnah, Moin and et al [126] introduce hybrid genetic algorithm with multiparents crossover for JSSP. Chakraborty1 and Bhowmik [127] propose an approach to JSSP using simulated annealing. Milad and Moslem and et al in [128] introduce an evolutionary approach for solving the JSSP in a service industry. Ponsich and coello [129] propose a hybrid differential evolution with TS algorithm for the solution of JSSP. Zhang, Song and Wu [130] introduce a hybrid artificial bee colony algorithm for JSSP.

Generally, the method for the JSSP mainly includes two kinds, one of which is exact methods and the other is approximation methods. Exact methods, such as branch and bound, linear programming and decomposition methods, guarantee global convergence and have been successful in solving small problems. Most scheduling problems in manufacturing systems are very complex to be solved by exact methods. It becomes increasingly important to explore ways of obtaining better schedules as priority dispatch, shifting bottleneck approach, local search, and heuristic methods.

In this chapter we propose a view of scheduling problems types. A structure of job shop scheduling problem and its formulation are proposed. The complexity of job shop scheduling problem is shown. Techniques for solving job shop scheduling are briefly introduced.

4.2 Scheduling Problems Types

The classical scheduling problems are classified in four distinct groups:

a) Workshops with one machine

There is only one machine which must be used for scheduling the given jobs, under the specified constraints.

B) Flow shop

There is more than one machine and each job must be processed on each of the machines. The number of operations for each job is equal with the number of machines. All jobs go through all the machines in the same order

C) Job shop

The problem is formulated under the same terms as for the flow shop problem. It differs in that each job has associated a processing order assigned for its operations.

D) Open shop

The same similarity with the flow shop problem, There are no ordering constraints on operations .The order for processing a job's operations is not relevant; any ordering will do.

4.3 Job shop scheduling problem structure

The JSSP will be described as following: for "n" jobs, each job needs several operations. These operations must be executed on "m" machines. The problem is to find out a schedule of the operations on the machines that minimizes the finishing time of the all the operations completed within the scheduled time. This finishing time is called as the makespan (C_{max}). This schedule should take into account some constraints. Generally, JSSP composed of the following elements [119]:

- **Jobs:** $j = \{j_1, j_2, ..., j_n\}$ is a set of n jobs to be scheduled. Each job j_i consists of a predetermined operation $O = \{O_{i1}, O_{i2}, ..., O_{ip}\}$ is the operation of j_i. All jobs are released at time 0.
- **Machines**: $m = \{m_1, m_2, ..., m_n\}$ is a set of m machines. Each machine can process only one operation at a time. Each operation can be processed without interruption during its performance on one of the set of machines. All machines are available at time 0.
- **Constraints**: the constraints that limit the possible assignments of the operations in JSSP can be divided mainly into following situations:
  - Disjunctive constraint : each operation can be performed by only one machine at a time
  - Non preemption constraint: each operation started should be completed.
  - Capacity constraint: operations are performed one after another on each machine.
  - Precedence/conjunctive constraint: for operations of different jobs, there are no precedence constraints among operations. For each job, each operation is forced to be scheduled after all predecessor operations.
  - Resource constraint: this constraint confirms that operations can be performed only by the machine from the given set [119].
- **Objective(s)**: most of the research reported in the literature treated with JSSP as single objective problem .This objective is to find a schedule that has minimum makespan. Some other researches treated with it as multiobjective problem. These

objectives are as flow time or tardiness. These objectives are also important like the makespan.

4.4 Job shop scheduling problem formulation

4.4.1 Mathematical representation of JSSP

Let the set of operations to be scheduled are $j = \{1, 2, ..., n, n+1\}$ and the set of machines are $M = \{1, 2, ..., m\}$. The operations 0 and $n+1$ are not original, and they have no duration and represent the initial and final operations $n+1$. Let the processing time of operation j is d_j [119]. Let the finish time of operation j is F_j. A schedule can be represented by a vector of finish times $\{f_1, f_2, ..., f_n\}$. Let $A(t)$ be the set of operations being processed at time t, and let $r_{j,m} = 1$ if operation j requires machine m to be processed otherwise $r_{j,m} = 0$. There are two types of constraints interrelated operations. First, the precedence constraints, each operation j is forced to be scheduled after all predecessor operations (p_j) are completed. Second, operation j can be scheduled only if the required machine is idle. The model of the JSSP can be described the following way:

$$\text{Minimize} \quad f_{n+1}(c_{\max}) \tag{4.1}$$

$$\text{subject to:} \quad f_k \le f_j - d_j \qquad j-1, ..., n+2 \ ; \ k \in p_j \tag{4.2}$$

$$\sum_{j \in A(t)} r_{j,m} \le 1 \qquad m \in M \ ; \ t \ge 0 \tag{4.3}$$

$$f_j \ge 0 \qquad j = 1, 2, ..., n+1 \tag{4.4}$$

Formula (4.1) is the objective of function that minimizes the finishing time of last Operation ($C_{\max}$). Constraints (4.2) indicate the precedence relations between operations for each job. Constraints (4.3) state that one machine can only process one operation at a time. Finally, constrain (4.4) forces the finishing time to be non-negative.

4.4.2 Disjunctive graph

Disjunctive graph formally described the JSSP. Disjunctive graph is presented by $G = (V; C \bigcup D)$; where V is a set of nodes representing operations of the jobs together with two special nodes, a source (0) and a sink (*), representing the beginning and end of the schedule, respectively. C is a set of conjunctive arcs representing technological sequence of the operations. D is a set of disjunctive arcs representing pairs of operations that must be performed on the same machines. As an example, table 4.1, 4.2 present data for (3 x 3) JSSP with three jobs and three machines. Table 4.1 includes the routing of each job through each machine and table 4.2 is the processing time for each operation. Figure 4.1 presents the disjunctive graph of this example [135].

Table 4.1: Machine sequence3 x 3 Example

Jobs	Machines		
J1	M1	M2	M3
J2	M1	M3	M2
J3	M2	M1	M3

Table 4.2: Processing time3 x3 Example

Jobs	Time		
J1	3	3	3
J2	2	3	4
J3	3	2	1

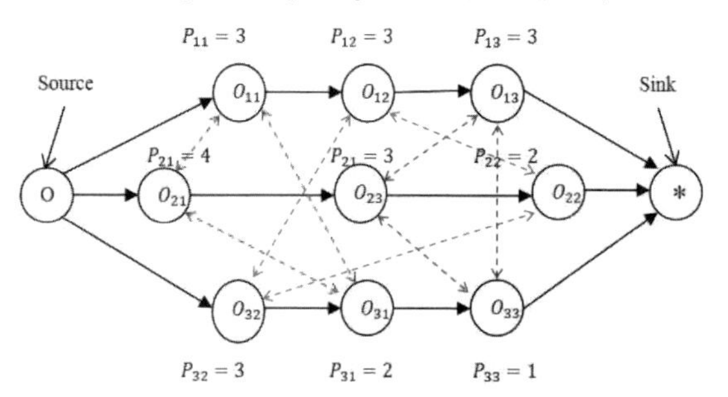

Figure 4.1 A disjunctive graph of a 3 x 3 Example

The solution of JSSP can defined as finding the ordering between all operations that must be processed on the same machine to fix precedence between these operations. In the disjunctive graph model, this is done by turning all undirected (disjunctive) arcs into directed ones. A selection is a set of directed arcs selected from disjunctive arcs. By definition, a selection is complete if all the disjunctions are selected. It is consistent if the resulting directed graph is a cyclic. The makespan is given by the length of the longest weighted path from source to sink in this graph.

4.4.3 Gantt-Chart

The Gantt-Chart is a convenient way of visually representing a solution of the JSSP. Gantt chart is adapted by Karol Adamiecki in 1896 [117] and independently by Henry Gantt in the 1910 that illustrates a project schedule. Gantt chart indicates the start and finish times of the terminal elements and the summary elements of a project. An example of a solution for the 3x3 problem in Tables 4.1 and 4.2 is given in Figure 4.2.

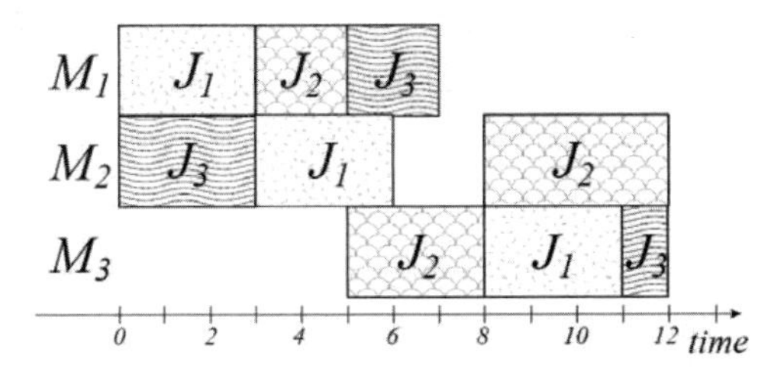

Figure 4.2 A Gantt-Chart representation of a solution for a 3 x3 Example

4.5 Complexity of JSSP

Practical experience appears that some computational problems are easier to solve than others. Complexity theory provides a mathematical framework in which computational problems are classified as easy or hard. Complexity theory divided problems into different classes according to the effort (runtime or memory) that is required to complete a globally best solution [120, 121]. It divided computational problem to P-problems and NP-problems.

1. In class of P-problems, the effort required to solve the problems increases with problem dimension by polynomial function. Consequently, the best solution for such problems can be obtained in most cases due to the computing power available today. It is even possible in the worst case just to try all possible solution candidates iteratively.
2. In class of NP-problems, in contrast, the effort required to solve the problems increases more than polynomial (i.e. exponential or even over-exponential). Increasing problem size increases the problem complexity. Therefore, NP-problems cannot be solved algorithmically even in rather low dimensions due to the enormous number of potential solution candidates that makes it impossible to find the global optimum in acceptable time.

In scheduling problems, since each job is partitioned into the operations processed on each machine, a schedule for a certain problem instance consists of an operation sequence for each machine involved. These operation sequences can be permuted independently from each other. There are a total of $(n!)^m$ different sequences. Therefore, scheduling problems are NP. Job Shop scheduling is a well known scheduling problem. JSSP is classified as NP-hard and one of the most intractable combinatorial optimization problems.

4.6 Job shop scheduling solving techniques

Scheduling problems are one of combinatorial Optimization (CO) problems. Combinatorial optimization is a topic that consists of finding an optimal object from a finite set of objects. In combinatorial optimization, the set of feasible solutions is discrete or can be reduced to discrete. According to the practical importance of CO problems, many methodologies to tackle them have been developed. These methodologies divided into two categories according to the obtained solution to exact or approximate methodologies.

Exact methodologies are guaranteed to find for finite size problem of a CO problem an optimal solution in bounded time. Yet, for CO problems that are NP hard, no polynomial time algorithm exists. Therefore, in the worst-case, exact methods might need exponential computation time. This often have too high computation times for practical purposes. Thus, the use of approximate methods has taken great attention in solving CO problems.

Approximate methods sacrifice the guarantee of finding optimal solutions for the sake of getting good solutions in a significantly reduced amount of time. As long as problems are small, exact techniques are slow and guarantee of global convergence. They involve mathematical programming, enumerate procedure and decomposition strategy. Approximate techniques are very fast but they do not guarantee for optimal solutions. Approximation methods involve constructive methods, insertion algorithms, evolutionary programs and local search techniques. Figure 4.3 shows the JSSP solving techniques [133,134].

Here we discuss a few of the most important contributions that solved JSSP .We take an evolutionary view of these techniques.

4.6.1 Exact techniques

4.6.1.1 Mathematical techniques

Mathematical programming has been applied extensively to JSSP. It has been formulated using integer programming, mixed-integer programming and dynamic programming. Until now, mathematical approaches have been limited because JSSP belongs to the class of NP-hard problems.

To overcome mathematical limitations, some researchers began to divide the scheduling problem into a number of sub problems. They propose number of techniques to solve them. Although, the computational power of modern computers have enabled these new techniques to solve larger problems. Still, difficulties have limited the use of these techniques [137].

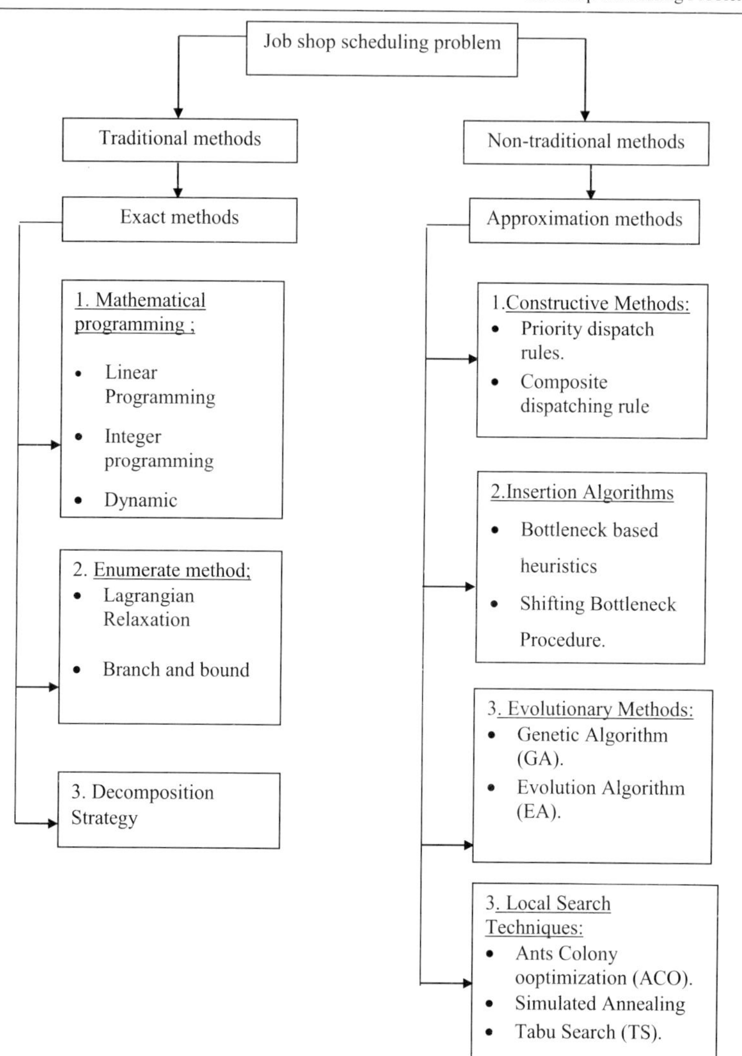

Figure 4.3 JSSP solving techniques

4.6.1.2 Enumerative techniques

The popular techniques of enumerative methods are branch-and-bound and Lagrangian relaxation. The branch-and-bound basic idea is that the problem is branched as a decision tree. Each node is a decision choice point and corresponds to partial solution. Although branch-and-bound procedures have been developed to speed up the search, large scheduling problems still have a very computational intensive procedure to be solved. Lagrangian relaxation technique solves the problem by skipping specific integer-valued constraints and then adds the corresponding costs to the objective function. Similar as branch and bound technique, Lagrangian relaxation technique has large limitations for large scheduling problems [136].

- **Branch and Bound**

In branch and bound technique, the solution space is represented by a dynamically constructed tree structure. Each node in the search tree represents a partial sequence of set of operations. Search begins at top node and then selects the next nodes. Selection is completed to the lowest level node. The branching operation starts from an unselected node and then determines the next set of possible nodes from which the search could progress. The bounding procedure selects the operation which will continue the search. Bounding procedure is based on estimation the lower bound and the best upper bound has been achieved. If the estimated lower bound at any node is greater than the current best upper bound, this partial selection and all its subsequent descendants are disregarded [134, 136].

4.6.1.3 Decomposition strategies

Decomposition strategies were proposed by Davis and Jones [137]. The decomposition was performed by reordering the constraints of the original problem to generate a block angular form. Then, this block angular form is transformed to a hierarchical tree structure. Generally, we would have N subproblems and a contained constraint as partial members of each subproblems. Latter, these constrains were termed the "coupling" constraints and included precedence relations and material handling. Decomposition strategies used two-levels. The top-level scheduler that is called as supremal determined the earliest start time and the latest finish time of each job. The lowest level scheduling modules that is called as infimals would refine these limit times for each job by detailed sequence of all operations. The supremal unit explicitly considered the coupling constraints, while the infimal units considered their individual decoupled constraint sets. Because of stochastic nature of job shops and its multiple and conflicting objective, it is difficult to express the coupling constraints using exact mathematical relationships. This made it almost impossible to develop a general solution methodology [137].

4.6.2 Approximate techniques

4.6.2.1 Constructive Methods

Two popular techniques of constructive methods are priority dispatch rules and composite dispatching rule. Dispatching rules have been consistently applied to scheduling problems. They are designed to have good solutions to complex problems in real-time. The term dispatching rule, scheduling rule, sequencing rule, or heuristic are often used synonymously. Dispatching rules are mainly classified according to the performance criteria for which they have been developed [134]. During the last 30 years, the performance of a large number of the dispatching rules has been extensively studied using simulation techniques. These studies have been aimed to answer the question - If you want to optimize a particular performance criterion, which rule should you choose? Most of the first work concentrated on the shortest processing time (SPT) rule. Recently, many investigations have been carried out to determine the dispatching rule. Dispatching rule optimizes a wide range of job-related as due date and tardiness and shop-related as throughput and utilization performance measures. The problem of selecting the best dispatching rule becomes a very active area of research.

4.6.2.2 Insertion Algorithms

Insertion algorithms are bottleneck heuristic and the shifting bottleneck heuristic. The shifting bottleneck heuristic is one of the powerful procedures among heuristics for the job shop scheduling problem. The idea is to solve for each machine a one-machine scheduling problem to optimality under the assumption that a lot of arc direct-shifting bottleneck heuristic, consists hof two subroutines. The first one (SB1) repeatedly solves one-machine scheduling problems while the second one (SB2) builds a partial enumeration. Each path from the root to a leaf is similar to an application of SB1. As the very name suggests, the shifting bottleneck heuristic always schedules bottleneck machines first. As a measure of the bottleneck quality of machine m, the value of an optimal solution of a certain one-machine scheduling problem on machine m is used. The one machine scheduling problems in consideration are those which arise from the disjunctive graph model when certain machines are already sequenced. The operation orders on sequenced machines are fully determined. Hence sequencing an additional machine probably results in a change of heads and tails of those operations whose machine order is still open. For all machines not sequenced, the maximum makespan of the corresponding optimal one-machine schedules, where the arc directions of the already sequenced machines are fixed, determines the bottleneck machine. In order to minimize the makespan of the job shop scheduling problem the bottleneck machine should be sequenced first [138].

4.6.2.3 Evolutionary Methods

EAs present a general purpose search procedures based on the mechanisms of natural selection and population genetics. EAs have been applied by many users in different areas of engineering, computer science, and operations research. Current evolutionary methods are evolutionary strategies, evolutionary programming, genetic programming and genetic algorithms.

Genetic algorithms inspired EAs. EAs are considered as a relatively new contribution to the field of artificial intelligence. They use various computational models of evolutionary processes to solve problems on a computer. EAs use a randomized but structured way to utilize genetic information to find a new search direction. They work by determining a goal in the form of a quality criterion and use this goal to compare solution candidates in a stepwise refinement of set data structures. If successful, after a number of iterations, an EA will return an optimal solution or a near-optimal one. The improvement process is achieved using genetic operators such as crossover and mutation. There are several variants of evolutionary algorithms, and also many hybrid systems which incorporate various features of these paradigms.

4.6.2.4. Local Search Techniques

In local search the generation mechanism outlines a neighbourhood for each configuration. A neighbourhood is a function which defines a simple transition from a solution to another solution by inducing a change that typically may be viewed as a small perturbation. Each new solution can be obtained directly from first solution by a single predefined partial transformation called a move. The final solution is conceived in the sense of one that satisfies certain structural requirements of a problem at hand, but it is not necessary neighborhoods.

- **Tabu search**

The basic idea of is that the search space of all feasible scheduling solutions is explored by a sequence of moves. The move from one schedule to another is made by evaluating all schedules and choosing the best one. Some moves may trap the search at a local optimum, or they lead to cycling (repeating part of the search). These moves are classified as tabu (i.e., they are forbidden) and are put onto something called the Tabu List. The tabu list is composed from the history of moves used during the search. These tabu moves force exploration of the search space until the old solution area (e.g., local optimum) is left behind. Another advantage in tabu search is freeing the search by a short term memory function that provides "strategic forgetting". Tabu search methods have been evolved to more advanced frameworks that include longer term memory mechanisms. These advanced frameworks are sometimes referred cas Adaptive Memory. Tabu search methods have been applied successfully to scheduling problems [39, 140].

CHAPTER 5
Hybrid Genetic Algorithm for Job Shop Scheduling Problems

5.1 Introduction

Scheduling has become a critical factor in many job shops especially for real-world industrial applications [115,116]. Finding the scheduling to achieve work in minimum time and more efficiently is called the job shop scheduling problem (JSSP). The JSSP is among the hardest combinatorial problems. Not only is complicated, but it is one of the worst NP-complete class members. In general, scheduling problems are NP; NP stands for non-deterministic polynomial, which means that it is not possible to solve an arbitrary problem in polynomial time. So, the JSSP has garnered attention due to both its practical importance and its solution complexity [118, 119].

The methods for the JSSP mainly include two kinds, one of which is exact methods and the other approximation methods [134]. In manufacturing systems, most scheduling problems are very complex and very complicated to be solved by exact methods. It becomes increasingly important to explore ways of obtaining better schedules. Recently, number of approximation techniques is used for JSSP such as genetic algorithm [141, 142], ant colony optimization [143,144], tabu search [145], particle swarm optimization [146-148], and Consultant Guided Search algorithm (CGS) [149].

GA has been used with increasing frequency to address scheduling problems. In [150] Lazár introduced a review of frequent approaches and methods for JSSP which most commonly are used in solving this problem. From this review we can say that GA is an effective metaheuristic to solve combinatorial optimization problems, and has been successfully adopted to solve the JSSP. How to adapt GAs to the JSSPs is very challenging but frustrating. Many efforts have been made in order to give an efficient implementation of GAs to the problem. In [151], a new GA is presented to solve the JSSP, while in [152] the impact of random initialization on solving the JSSP is addressed and using GA as an optimization technique.

Due to the complexity of the JSSP, using simple GA to solve this problem may not be more efficient in practice. Much effort in the literature has focused on hybrid methods [140-156]. Qing and Wing [154] solve JSSP more effectively through designed some GA operators (mixed selection operator, new crossover operator and mutation operator based on the critical path). In [155] Spanos and Ponis proposed a new hybrid parallel GA with specialized crossover and mutation operators utilizing path-relinking concepts from combinatorial optimization approaches. Some researchers focused their attention on combining the GA with local search schemes to develop some hybrid optimization strategies for JSSP

In this chapter, we propose a new hybrid algorithm for JSSP based on GA and local search. Firstly, we design a GA model for JSSP; where a new initialization technique, modified crossover and mutation operators is proposed. Finally, local search based on the neighborhood structure is applied to GA result to improve the solution quality. The proposed approach is tested by a set of standard instances taken from the literature [157,158].

5.2 The proposed algorithm (HGA)

The proposed algorithm is a combination between genetic algorithm and the local search. We start by a generation alternation model using genetic algorithm [148]. A new way of initializing the population of solutions is designed. Advanced crossover and mutation operators are used to improve the solution quality. Then local search based on the neighborhood is applied in GA result. The approach is tested on a set of standard instances taken from the OR-library [157,158]. The computation results have validated the effectiveness of the proposed algorithm.

5.2.1 Phase I : GA

Step 1: Initial population

For JSSP, the initial solution plays a critical role in determining the quality of final solution. However, the initial population has been produced randomly, it is not efficient. It will require longer search time to obtain an optimal solution and also decreases the search possibility for an optimal solution giving infeasible solutions [159-165].

In this chapter, a new way of initializing first population of solutions is designed. This new method gives feasible solutions and is very useful in large problem. We generate the initial population (job sequence in machines) by taking into account machine sequence for jobs. For example, problem in Figure 4.1, for machine 1 there are two jobs (1,2) should start in it according to their machine sequence. In our approach we take this in account. Start in machine 1 with one of these jobs. Then complete jobs sequence in machine 1 with same manner. Figure 5.1 illustrates this method in the disjunctive graph; where jobs in machine are ordering with the earliest jobs from start point (O). Then the job sequence in machine 1 will be (1,2,3), (2,1,3). Use same manner for rest machines.

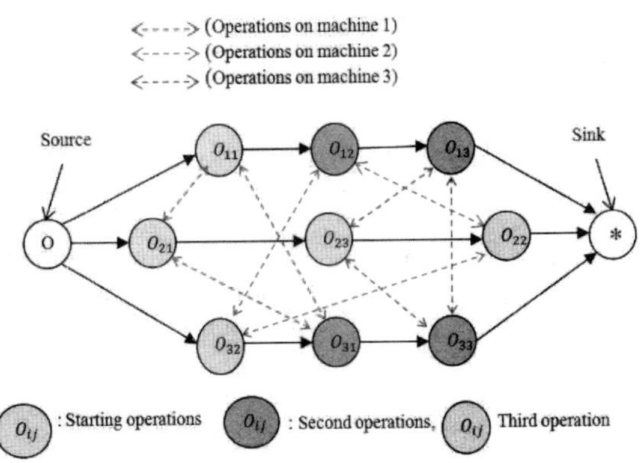

Figure 5.1: Illustration of initialization technique with disjunctive graph for a 3 x 3 example

Step 2: Evaluation

In JSSP, makespan represents a good performance measure. Makespan of the jobs (Cmax) is the completion time of all the jobs where, the schedule with the minimal makespan often implies a high utilization of machines. The makespan objective is selected for comparison and assessing the performance of the generated solutions by genetic algorithm.

Step 3: Create a new population

Create a new population from the current population by applying the genetic search operators (selection, crossover, mutation) one after another [161]. First, individuals are selected in a competitive manner, based on their fitness as measured by the objective function (their makespan). Then the solutions are ranked according to fitness function. After that, other genetic search operators such as mutation and crossover are then applied to obtain a new generation of chromosomes. Expected quality of new generation over all the chromosomes is better than that of the previous generation.

a) Selection

For JSSP as any optimization problem there are several techniques of selection such as, steady state selection, stochastic universal sampling, etc. Here we used roulette wheel selection. In roulette wheel, the selection is according to chromosome's fitness. Better chromosomes are selected as parents. Each individual is represented by a slice of circular roulette wheel, and the size of slice is proportional to the individual fitness of chromosomes. The Roulette wheel algorithm starts with calculating the sum of all chromosomes fitness in the population and every chromosome's probability according to equation 2.9. Then generate random number between zero and one. Finally, go through the chromosomes' probability. When the generated number is more than a chromosome's probability, stop and select this chromosome.

b) Crossover

The crossover operates on two chromosomes at a time and generates offspring by combining features of both chromosomes. In general, the performance of the genetic algorithms depends, to a great extent, on the performance of the crossover operator used. During the past two decades, various crossover operators have been proposed for literal permutation encodings for JSSP, such as partial-mapped crossover (PMX), order crossover (OX), cycle crossover (CX), position-based crossover, order-based crossover, etc. [153].

In this chapter, we use advance OX that proposed in [165]. Gao and Sun applying OX in operation sequence, here we advanced it by applying OX in job sequence for every machine. Advanced order crossover can be described as follows:

- Select a subsection of job sequence for a machine from one parent at random.
- Produce offspring by copying the substring of job sequence for a machine into the corresponding positions.
- Delete the jobs that are already in the substring from the second parent. The resulted sequence of jobs contains jobs that the offspring needs.
- Place the jobs into the unfixed positions of the offspring from left to right according to the order of the sequence in the second parent.

The procedure is illustrated in Figure 5.2.

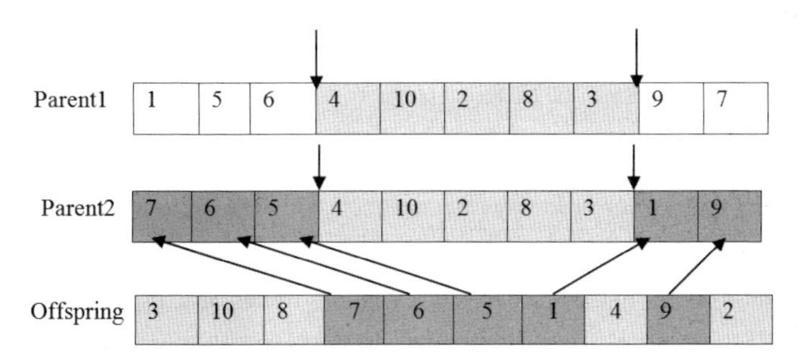

Figure 5.2 Order crossover on job sequence for one machine

c) Mutation

Mutation is just used to produce small perturbations on chromosomes in order to maintain the diversity of population. During the last decade, there are several mutation operators have been proposed such as inversion, insertion, displacement, reciprocal exchange mutation, and shift mutation [161]. Inversion mutation selects two positions within a chromosome at random and then inverts the substring between these two positions. In this paper we advanced inversion mutation, by choosing two or more positions in the jobs sequence for one machine and invert them as showed in Figure 5.3 then apply the same steps for other machines, we get a new offspring.

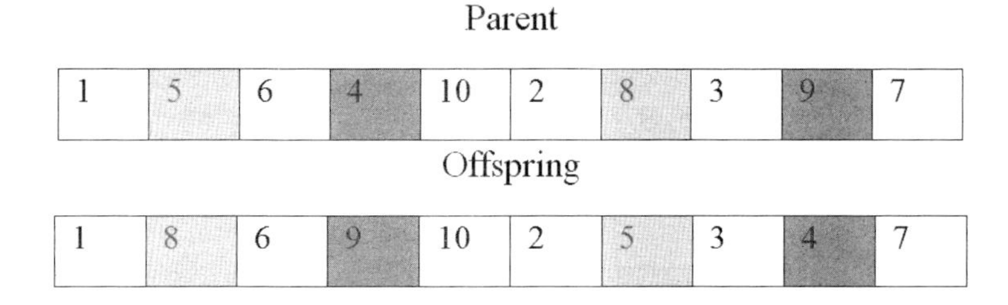

Figure 5.3 Inversion mutation on job sequence for one machine

Step 4: Migration

In this step, the new generation composed by migration initial population with and offspring population. We have taken the best individuals of initial population with generation gap percentage and offspring population [163].

Step 5: Termination test

The algorithm is terminated when the maximum number of generations is achieved, or when the individuals of the population converges, convergence occur when all individuals positions in the population are identical. In this case, crossover will have no further effect. Otherwise, create other new population.

5.2.2 Phase II: Local search

Using genetic algorithms for the JSSP are usually with a slow convergence speed and easy to trap into local optimal solutions. In order to enhance the convergence speed, we combine the genetic algorithm with local search schemes to develop some hybrid optimization strategies for JSSP improving the solution quality.

We use the approach of Nowicki and Smutnicki (1996) [156, 157]; where the local search procedure starts by finding the critical path for the solution obtained by GA. The critical path decomposes by a number of blocks where a block is a maximal sequence of adjacent operations that require the same machine. Any operation on the critical path is called a critical operation. Critical path can be determined in the Gantt-Chart representation for the solution that it is the longest continuous path in the Gantt-Chart representation. Then, a neighborhood is defined as interchanges of any two sequential operations in the critical path. Interchanging the last two or the first two critical operations gives better solution. The local search procedure is shown in Figure 5.4, while Figure 5.5 shows the flow chart of the proposed algorithm (HGA).

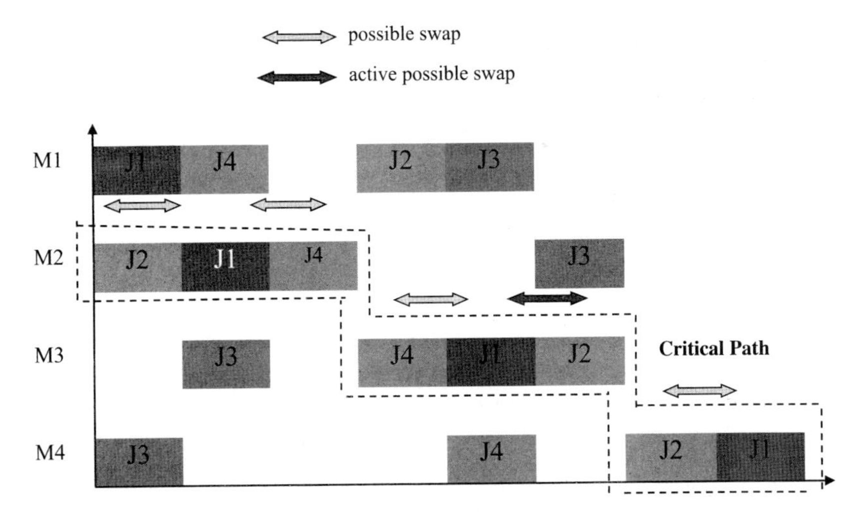

Figure 5.4 The local search procedure.

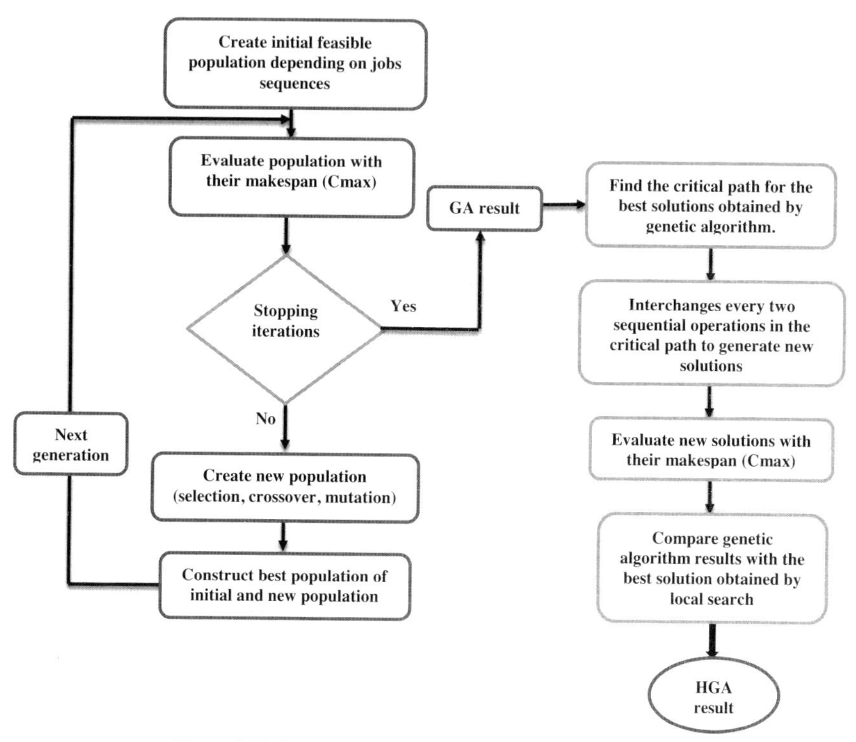

Figure 5.5 HGA procedures for Job shop scheduling problems

5.3 Experimental Results

To illustrate the effectiveness of the algorithm described in this chapter. We consider some standard JSSP test problems taken from the literature. The performance comparison with other optimization algorithm was done in order to demonstrate the efficiency and robustness of the proposed algorithm. The algorithm is coded in MATLAB 7.8 and the simulations have been executed on an Intel core (TM) i7-4510u cpu@ 2.00GHZ 2.60 GHz processor. The proposed algorithm, involves a number of parameters that affect the performance of algorithm. The parameters are summarized in Table 5.1.

5.3.1 Test Problems

To illustrate the effectiveness of the algorithm described in this chapter, we consider some standard JSSP test problems: Fisher and Thompson (1963) [157] instances FT06 and Lawrence (1984) instances LA01 to LA20 [158]. Tables 5.2-5.22 propose the data for the test function.

Table 5.1: The proposed algorithm parameters

Generation gap	0.9
Crossover rate	0.9
Mutation rate	0.7
Selection operator	roulette wheel selection Single point
Crossover operator	order crossover
Mutation operator	Inversion mutation
GA generation	50-1000

Table 5.2: The machines sequence and processing time for the problem FT06 (6x6)

Jobs	Machines sequence						Processing Time					
J1	3	1	2	4	6	5	1	3	6	7	3	6
J2	2	3	5	6	1	4	8	5	10	10	10	4
J3	3	4	6	1	2	5	5	4	8	9	1	7
J4	2	1	3	4	5	6	5	5	5	3	8	9
J5	3	2	5	6	1	4	9	3	5	4	3	1
J6	2	4	6	1	5	3	3	3	9	10	4	1

Table 5.3: The machines sequence and processing time for the problem LA01 (10x5)

Jobs	Machines sequence					Processing Time				
J1	2	1	5	4	3	21	53	95	55	34
J2	1	4	5	3	2	21	52	16	26	71
J3	4	5	2	3	1	39	98	42	31	12
J4	2	1	5	3	4	77	55	79	66	77
J5	1	4	3	2	5	83	34	64	19	37
J6	2	3	5	1	4	54	43	79	92	62
J7	4	5	2	3	1	69	77	87	87	93
J8	3	1	2	4	5	38	60	41	24	83
J9	4	2	5	1	3	17	49	25	44	98
J10	5	4	3	2	1	77	79	43	75	96

Table 5.4: The machines sequence and processing time for the problem LA02 (10x5)

Jobs	Machines sequence					Processing Time				
J1	1	4	2	5	3	20	87	31	76	17
J2	5	3	1	2	4	25	32	24	18	81
J3	2	3	5	1	4	72	23	28	58	99
J4	3	2	5	1	4	86	76	97	45	90
J5	5	1	4	3	2	27	42	48	17	46
J6	2	1	5	4	3	67	98	48	27	62
J7	5	2	4	1	3	28	12	19	80	50
J8	2	1	3	4	5	63	94	98	50	80
J9	5	1	3	2	4	14	75	50	41	55
J10	5	3	2	4	1	72	18	37	79	61

Table 5.5: The machines sequence and processing time for the problem LA03 (10x5)

Jobs	Machines sequence					Processing Time				
J1	2	3	1	5	4	23	45	82	84	38
J2	3	2	1	5	4	21	29	18	41	50
J3	3	4	5	1	2	38	54	16	52	52
J4	5	1	3	2	4	37	54	74	62	57
J5	5	1	2	4	3	57	81	61	68	30
J6	5	1	2	3	4	81	79	89	89	11
J7	4	3	1	5	2	33	20	91	20	66
J8	5	2	1	3	4	24	84	32	55	8
J9	5	1	4	3	2	56	7	54	64	39
J10	5	2	1	3	4	40	83	19	8	7

Table 5.6: The machines sequence and processing time for the problem LA04 (10x5)

Jobs	Machines sequence					Processing Time				
J1	1	3	4	5	2	12	94	92	91	7
J2	2	4	5	3	1	19	11	66	21	87
J3	2	1	4	5	3	14	75	13	16	20
J4	3	5	1	4	2	95	66	7	7	77
J5	2	4	5	1	3	45	6	89	15	34
J6	4	3	1	5	2	77	20	76	88	53
J7	3	2	1	4	5	74	88	52	27	9
J8	2	4	1	5	3	88	69	62	98	52
J9	3	5	1	2	4	61	9	62	52	90
J10	3	5	4	2	1	54	5	59	15	88

Table 5.7: The machines sequence and processing time for the problem LA05 (10x5)

Jobs	Machines sequence					Processing Time				
J1	2	1	5	3	4	72	87	95	66	60
J2	5	4	1	3	2	5	35	48	39	54
J3	2	4	3	1	5	46	20	21	97	55
J4	1	4	5	2	3	59	19	46	34	37
J5	5	3	4	2	1	23	73	25	24	28
J6	4	1	5	2	3	28	45	5	78	83
J7	1	4	2	5	3	53	71	37	29	12
J8	5	3	4	2	1	12	87	33	55	38
J9	3	4	2	1	5	49	83	40	48	7
J10	3	4	1	5	2	65	17	90	27	23

Table 5.8: The machines sequence and processing time for the problem LA06 (15x5)

Jobs	Machines sequence					Processing Time				
J1	2	3	5	1	4	21	34	95	53	55
J2	4	5	2	3	1	52	16	71	26	21
J3	3	1	2	4	5	31	12	42	39	98
J4	4	2	5	1	3	77	77	79	55	66
J5	5	4	3	2	1	37	34	64	19	83
J6	3	2	1	4	5	43	54	92	62	79
J7	1	4	2	5	3	93	69	87	77	87
J8	1	2	3	5	4	60	41	38	83	24
J9	3	4	5	1	2	98	17	25	44	49
J10	1	5	4	2	3	96	77	79	75	43
J11	5	3	1	4	2	28	35	95	76	7
J12	1	5	3	2	4	61	10	95	9	35
J13	5	4	2	3	1	59	16	91	59	46
J14	5	2	1	3	4	43	52	28	27	50
J15	1	2	3	5	4	87	45	39	9	41

Table 5.9: The machines sequence and processing time for the problem LA07 (15x5)

Jobs	Machines sequence					Processing Time				
J1	1	5	2	4	3	47	57	71	96	14
J2	1	2	5	4	3	75	60	22	79	65
J3	4	1	3	2	5	32	33	69	31	58
J4	1	2	5	4	3	44	34	51	58	47
J5	4	2	1	3	5	29	44	62	17	8
J6	2	3	1	5	4	15	40	97	38	66
J7	3	2	1	5	4	58	39	57	20	50
J8	3	4	5	1	2	57	32	87	63	21
J9	5	1	3	2	4	56	84	90	85	61
J10	5	1	2	4	3	15	20	67	30	70
J11	5	1	2	3	4	84	82	23	45	38

Table 5.9 (continued)

Jobs	Machines sequence					Processing Time				
J12	4	3	1	5	2	50	21	18	41	29
J13	5	2	1	3	4	16	52	52	38	54
J14	5	1	4	3	2	37	54	57	74	62
J15	5	2	1	3	4	57	61	81	30	68

Table 5.10: The machines sequence and processing time for the problem LA08 (15x5)

Jobs	Machines sequence					Processing Time				
J1	4	3	1	5	2	92	94	12	91	7
J2	3	2	1	4	5	21	19	87	11	66
J3	2	4	1	5	3	14	13	75	16	20
J4	3	5	1	2	4	95	66	7	77	7
J5	3	5	4	2	1	34	89	6	45	15
J6	5	4	3	2	1	88	77	20	53	76
J7	5	4	1	2	3	9	27	52	88	74
J8	4	3	1	2	5	69	52	62	88	98
J9	4	1	5	3	2	90	62	9	61	52
J10	5	3	4	1	2	5	54	59	88	15
J11	1	2	5	4	3	41	50	78	53	23
J12	1	5	3	4	2	38	72	91	68	71
J13	1	4	5	3	2	45	95	52	25	6
J14	4	2	1	5	3	30	66	23	36	17
J15	3	1	4	2	5	95	71	76	8	88

Table 5.11: The machines sequence and processing time for the problem LA09 (15x5)

Jobs	Machines sequence					Processing Time				
J1	2	4	3	1	5	66	85	84	62	19
J2	4	2	3	5	1	59	64	46	13	25
J3	5	4	2	3	1	88	80	73	53	41
J4	1	2	3	4	5	14	67	57	74	47
J5	1	5	3	4	2	84	64	41	84	78
J6	1	4	2	3	5	63	28	46	26	52
J7	4	3	5	2	1	10	17	73	11	64
J8	3	2	4	5	1	67	97	95	38	85
J9	3	5	1	2	4	95	46	59	65	93
J10	3	5	4	2	1	43	85	32	85	60
J11	5	4	3	1	2	49	41	61	66	90
J12	2	1	4	5	3	17	23	70	99	49
J13	5	4	1	2	3	40	73	73	98	68
J14	4	2	3	1	5	57	9	7	13	98
J15	1	2	3	5	4	37	85	17	79	41

Table 5.12: The machines sequence and processing time for the problem LA10 (15x5)

Jobs	Machines sequence					Processing Time				
J1	2	3	4	1	5	58	44	5	9	58
J2	2	1	5	4	3	89	97	96	77	84
J3	1	2	3	5	4	77	87	81	39	85
J4	4	2	3	1	5	57	21	31	15	73
J5	3	1	2	4	5	48	40	49	70	71
J6	4	5	3	1	2	34	82	80	10	22
J7	2	5	1	3	4	91	75	55	17	7
J8	3	4	2	5	1	62	47	72	35	11
J9	1	4	5	2	3	64	75	50	90	94
J10	3	5	4	1	2	67	20	15	12	71
J11	1	5	4	3	2	52	93	68	29	57
J12	3	1	2	5	4	70	58	93	7	77
J13	4	3	2	5	1	27	82	63	6	95
J14	2	3	5	1	4	87	56	36	26	48
J15	4	3	1	5	2	76	36	36	15	8

Table 5.13: The machines sequence and processing time for the problem LA11 (20x5)

Jobs	Machines sequence					Processing Time				
J1	3	2	1	4	5	34	21	53	55	95
J2	1	4	2	5	3	21	52	71	16	26
J3	1	2	3	5	4	12	42	31	98	39
J4	3	4	5	1	2	66	77	79	55	77
J5	1	5	4	2	3	83	37	34	19	64
J6	5	3	1	4	2	79	43	92	62	54
J7	1	5	3	2	4	93	77	87	87	69
J8	5	4	2	3	1	83	24	41	38	60
J9	5	2	1	3	4	25	49	44	98	17
J10	1	2	3	5	4	96	75	43	77	79
J11	1	4	2	5	3	95	76	7	28	35
J12	5	3	1	2	4	10	95	61	9	35
J13	2	3	5	1	4	91	59	59	46	16
J14	3	2	5	1	4	27	52	43	28	50
J15	5	1	4	3	2	9	87	41	39	45
J16	2	1	5	4	3	54	20	43	14	71
J17	5	2	4	1	3	33	28	26	78	37
J18	2	1	3	4	5	89	33	8	66	42
J19	5	1	3	2	4	84	69	94	74	27
J20	5	3	2	4	1	81	45	78	69	96

Table 5.14: The machines sequence and processing time for the problem LA12 (20x5)

Jobs	Machines sequence					Processing Time				
J1	2	1	5	3	4	23	82	84	45	38
J2	4	5	2	1	3	50	41	29	18	21
J3	5	4	2	3	1	16	54	52	38	52

Table 5.14 (continued)

Jobs	Machines sequence					Processing Time				
J4	2	4	5	3	1	62	57	37	74	54
J5	4	2	3	1	5	68	61	30	81	57
J6	2	3	4	1	5	89	89	11	79	81
J7	2	1	4	5	3	66	91	33	20	20
J8	4	5	3	1	2	8	24	55	32	84
J9	1	3	2	5	4	7	64	39	56	54
J10	1	5	4	3	2	19	40	7	8	83
J11	1	3	4	5	2	63	64	91	40	6
J12	2	4	5	3	1	42	61	15	98	74
J13	2	1	4	5	3	80	26	75	6	87
J14	3	5	1	4	2	39	22	75	24	44
J15	2	4	5	1	3	15	79	8	12	20
J16	4	3	1	5	2	26	43	80	22	61
J17	3	2	1	4	5	62	36	63	96	40
J18	2	4	1	5	3	33	18	22	5	10
J19	3	5	1	2	4	64	64	89	96	95
J20	3	5	4	2	1	18	23	15	38	8

Table 5.15: The machines sequence and processing time for the problem LA13 (20x5)

Jobs	Machines sequence					Processing Time				
J1	4	1	2	5	3	60	87	72	95	66
J2	2	1	3	4	5	54	48	39	35	5
J3	4	2	1	3	5	20	46	97	21	55
J4	3	1	4	2	5	37	59	19	34	46
J5	3	4	2	1	5	73	25	24	28	23
J6	2	4	3	1	5	78	28	83	45	5
J7	4	2	3	5	1	71	37	12	29	53
J8	5	4	2	3	1	12	33	55	87	38
J9	1	2	3	4	5	48	40	49	83	7
J10	1	5	3	4	2	90	27	65	17	23
J11	1	4	2	3	5	62	85	66	84	19
J12	4	3	5	2	1	59	46	13	64	25
J13	3	2	4	5	1	53	73	80	88	41
J14	3	5	1	2	4	57	47	14	67	74
J15	3	5	4	2	1	41	64	84	78	84
J16	5	4	3	1	2	52	28	26	63	46
J17	2	1	4	5	3	11	64	10	73	17
J18	5	4	1	2	3	38	95	85	97	67
J19	4	2	3	1	5	93	65	95	59	46
J20	1	2	3	5	4	60	85	43	85	32

Table 5.16: The machines sequence and processing time for the problem LA14 (20x5)

Jobs	Machines sequence					Processing Time				
J1	2	5	1	3	4	5	58	44	9	58
J2	3	4	2	5	1	89	96	97	84	77
J3	1	4	5	2	3	81	85	87	39	77
J4	3	5	4	1	2	15	57	73	21	31
J5	1	5	4	3	2	48	71	70	40	49
J6	3	1	2	5	4	10	82	34	80	22
J7	4	3	2	5	1	17	55	91	75	7
J8	2	3	5	1	4	47	62	72	35	11
J9	4	3	1	5	2	90	94	50	64	75
J10	5	3	1	2	4	15	67	12	20	71
J11	4	2	1	3	5	93	29	52	57	68
J12	2	4	1	5	3	77	93	58	70	7
J13	5	1	4	3	2	63	27	95	6	82
J14	3	2	5	4	1	36	26	48	56	87
J15	5	2	4	1	3	36	8	15	76	36
J16	2	1	5	4	3	78	84	41	30	76
J17	1	5	3	2	4	78	75	88	13	81
J18	2	5	1	4	3	54	40	13	82	29
J19	4	2	1	3	5	26	82	52	6	6
J20	2	5	1	3	4	54	64	54	32	882

Table 5.17: The machines sequence and processing time for the problem LA15 (20x5)

Jobs	Machines sequence					Processing Time				
J1	1	3	2	4	5	6	40	81	37	19
J2	3	4	1	5	2	40	32	55	81	9
J3	2	5	3	4	1	46	65	70	55	77
J4	3	5	1	4	2	21	65	64	25	15
J5	3	1	2	4	5	85	40	44	24	37
J6	1	5	2	4	3	89	29	83	31	84
J7	5	4	2	3	1	59	38	80	30	8
J8	1	3	2	5	4	80	56	77	41	97
J9	5	1	4	3	2	56	91	50	71	17
J10	2	1	5	3	4	40	88	59	7	80
J11	1	2	3	5	4	45	29	8	77	58
J12	3	1	4	2	5	36	54	96	9	10
J13	1	3	2	4	5	28	73	98	92	87
J14	1	4	3	2	5	70	86	27	99	96
J15	2	1	5	4	3	95	59	56	85	41
J16	2	3	5	1	4	81	92	32	52	39
J17	2	5	3	1	4	7	22	12	88	60
J18	4	1	3	5	2	45	93	69	49	27
J19	1	2	3	4	5	21	84	61	68	26
J20	2	3	5	1	4	82	33	71	99	44

Table 5.18: The machines sequence and processing time for the problem LA16 (20x5)

Jobs	Machines sequence										Processing Time										
J1	2	7	10	9	8	3	1	5	4	6	21	71	16	52	26	34	53	21	55	95	21
J2	5	3	6	10	1	8	2	9	7	4	55	31	98	79	12	66	42	77	77	39	55
J3	4	3	9	2	5	10	8	7	1	6	34	64	62	19	92	79	43	54	83	37	34
J4	2	4	3	8	9	10	7	1	6	5	87	69	87	38	24	83	41	93	77	60	87
J5	3	1	6	7	8	2	5	10	4	9	98	44	25	75	43	49	96	77	17	79	98
J6	3	4	6	10	5	7	1	9	2	8	35	76	28	10	61	9	95	35	7	95	35
J7	4	3	1	2	10	9	7	6	5	8	16	59	46	91	43	50	52	59	28	27	16
J8	2	1	4	5	7	10	9	6	3	8	45	87	41	20	54	43	14	9	39	71	45
J9	5	3	9	6	4	8	2	7	10	1	33	37	66	33	26	8	28	89	42	78	33
J10	9	10	3	5	4	1	8	7	2	6	69	81	94	96	27	69	45	78	74	84	69

Table 5.19: The machines sequence and processing time for the problem LA17 (10x10)

Jobs	Machines sequence										Processing Time										
J1	5	8	10	3	4	9	6	7	2	1	18	21	41	45	38	50	84	29	23	82	18
J2	9	6	2	8	3	4	7	10	5	1	57	16	52	74	38	54	62	37	54	52	57
J3	3	5	4	2	9	7	8	1	10	6	30	79	68	61	11	89	89	81	81	57	30
J4	1	9	4	8	6	3	5	7	2	10	91	8	33	55	20	20	32	84	66	24	91
J5	10	1	5	9	7	3	6	4	8	2	40	7	19	7	83	64	56	54	8	39	40
J6	4	3	6	1	8	5	9	2	7	10	91	64	40	63	98	74	61	6	42	15	91
J7	2	8	9	4	5	6	7	1	3	10	80	39	24	75	75	6	44	26	87	22	80
J8	2	8	3	1	9	7	4	10	6	5	15	43	20	12	26	61	79	22	8	80	15
J9	3	4	5	10	1	7	8	9	2	6	62	96	22	5	63	33	10	18	36	40	62
J10	2	1	6	4	10	8	9	3	7	5	96	89	64	95	23	18	15	64	38	8	96

Table 5.20: The machines sequence and processing time for the problem LA18 (10x10)

Jobs	Machines sequence										Processing Time										
J1	2	7	10	9	8	3	1	5	4	6	21	71	16	52	26	34	53	21	55	95	21
J2	5	3	6	10	1	8	2	9	7	4	55	31	98	79	12	66	42	77	77	39	55
J3	4	3	9	2	5	10	8	7	1	6	34	64	62	19	92	79	43	54	83	37	34
J4	2	4	3	8	9	10	7	1	6	5	87	69	87	38	24	83	41	93	77	60	87
J5	3	1	6	7	8	2	5	10	4	9	98	44	25	75	43	49	96	77	17	79	98
J6	3	4	6	10	5	7	1	9	2	8	35	76	28	10	61	9	95	35	7	95	35
J7	4	3	1	2	10	9	7	6	5	8	16	59	46	91	43	50	52	59	28	27	16
J8	2	1	4	5	7	10	9	6	3	8	45	87	41	20	54	43	14	9	39	71	45
J9	5	3	9	6	4	8	2	7	10	1	33	37	66	33	26	8	28	89	42	78	33
J10	9	10	3	5	4	1	8	7	2	6	69	81	94	96	27	69	45	78	74	84	69

Table 5.21: The machines sequence and processing time for the problem LA19 (10x10)

Jobs	Machines sequence										Processing Time										
1	7	1	5	4	8	9	2	6	3	10	54	87	48	60	39	35	72	95	66	5	54
J2	4	10	7	6	1	9	5	3	8	2	20	46	34	55	97	19	59	21	37	46	20
J3	5	2	9	1	8	7	6	4	10	3	45	24	28	28	83	78	23	25	5	73	45
J4	10	2	5	4	9	3	7	1	8	6	12	37	38	71	33	12	55	53	87	29	12
J5	4	3	7	10	8	1	5	6	2	9	83	49	23	27	65	48	90	7	40	17	83
J6	2	5	1	3	10	7	8	9	6	4	66	25	62	84	13	64	46	59	19	85	66
J7	2	4	1	3	10	8	9	5	7	6	73	80	41	53	47	57	74	14	67	88	73
J8	6	4	7	2	1	8	9	10	3	5	64	84	46	78	84	26	28	52	41	63	64
J9	2	1	8	5	4	6	10	9	7	3	11	64	67	85	10	73	38	95	97	17	11
J10	5	9	3	4	2	7	8	10	6	1	60	32	95	93	65	85	43	85	46	59	60

Table 5.22: The machines sequence and processing time for the problem LA20 (10x10)

Jobs	Machines sequence										Processing Time										
J1	3	4	6	5	1	8	9	10	2	7	44	5	58	97	9	84	77	96	58	89	44
J2	5	8	2	9	1	4	3	6	10	7	15	31	87	57	77	85	81	39	73	21	15
J3	10	7	5	4	2	1	9	3	8	6	82	22	10	70	49	40	34	48	80	71	82
J4	2	3	8	6	9	5	4	7	10	1	91	17	62	75	47	11	7	72	35	55	91
J5	7	2	4	1	3	9	5	8	10	6	71	90	75	64	94	15	12	67	20	50	71
J6	8	6	9	3	5	7	4	2	10	1	70	93	77	29	58	93	68	57	7	52	70
J7	7	2	5	6	3	4	8	9	10	1	87	63	26	6	82	27	56	48	36	95	87
J8	1	6	9	10	4	7	5	8	3	2	36	15	41	78	76	84	30	76	36	8	36
J9	6	3	4	7	5	8	9	10	2	1	88	81	13	82	54	13	29	40	78	75	88
J10	10	5	7	8	1	3	9	6	4	2	88	54	64	32	52	6	54	82	6	26	88

5.3.2 Results and discussion

Table 5.23 shows the experimental results. It lists problem name, problem size (number of jobs ×number of operations), problem complexity indicated by the number of probable solutions that the algorithm search in, the best known solution (BKS) for the test problems, solution obtained by applying our genetic algorithm only, the results obtained by hybridizing the local search with genetic algorithm (HGA) and the improvement percentage in the obtained results as due to hybrid local search with genetic algorithm.

From the results in Table 5.23, hybridizing the local search with genetic algorithm improve the obtained results. The mean improvement percentage is about 5.36%. Figure 5.6 showed the improvement percentage due to hybrid local search with genetic algorithm and using local search with GA has better performance than GA only and it is possible to accelerate convergence speed, good quality of optimal solutions and more predictable application behavior.

Table 5.23: The makespan comparison between BKS in literature works, and the proposed algorithm results (GA and HGA)

Problem (size)	Number of probable solution= (n!)^m	Best Known Solution (BKS)	GA	HGA	Improv. Percent.
Ft06 (6×6)	1.4413e+019	55	55	55	0.000%
LA01 (10×5)	6.2924e+032	666	666	666	0.000%
LA02 (10×5)	6.2924e+032	655	790	714	11.603%
LA03 (10×5)	6.2924e+032	597	683	617	11.055%
LA04 (10×5)	6.2924e+032	590	672	632	6.779%
LA05 (10×5)	6.2924e+032	593	593	593	0.000%
LA06 (15×5)	3.8238e+060	926	926	926	0.000%
LA07 (15×5)	3.8238e+060	890	916	899	1.91%
LA08 (15×5)	3.8238e+060	863	863	863	0.000%
LA09 (15×5)	3.8238e+060	951	951	951	0.000%
LA10 (15×5)	3.8238e+060	958	958	958	0.000%
LA11 (20×5)	8.5236e+091	1222	1222	1222	0.000%
LA12 (20×5)	8.5236e+091	1039	1039	1039	0.000%
LA13 (20×5)	8.5236e+091	1150	1150	1150	0.000%
LA14 (20×5)	8.5236e+091	1292	1292	1292	0.000%
LA15 (20×5)	8.5236e+091	1207	1324	1311	1.077%
LA16 (10×10)	3.9594e+065	945	1140	1077	6.667%
LA17 (10×10)	3.9594e+065	784	848	843	0.638%
LA18 (10×10)	3.9594e+065	848	948	894	6.368%
LA19 (10×10)	3.9594e+065	842	961	896	7.719%
LA20 (10×10)	3.9594e+065	902	956	949	0.776%

In addition, we propose some obtained optimal solution by our hybrid algorithm problems. The solutions of problems (FT06-LA01-LA05) are proposed in Tables 5.24-5.26. The tables illustrated the job sequence on every machine and the starting time for every job on every machine and the finish time for every job on every machine tables. The Gantt charts for the problems are proposed in Figures 5.7-5.9.

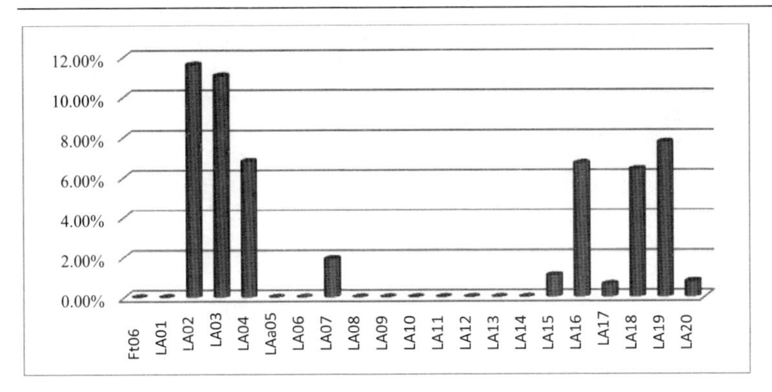

Figure 5.6 The improvement percentage due to hybrid of local search with GA.

Table 5.24: The solution of problem FT06

Job Sequence						Start Time						Finish Time					
M1	M2	M3	M4	M5	M6	M1	M2	M3	M4	M5	M6	M1	M2	M3	M4	M5	M6
3	2	3	3	0	1	9	13	5	0	8	6	17	23	9	5	17	9
6	5	5	1	8	4	19	25	16	5	13	13	28	30	19	6	28	18
2	3	4	2	13	3	28	30	27	8	16	18	38	37	30	13	38	27
5	4	1	5	22	6	38	37	34	13	25	28	42	45	41	22	42	38
1	6	2	4	27	2	42	45	48	22	28	38	45	49	52	27	45	48
4	1	5	6	0	5	45	49	52	49	34	48	54	55	53	50	54	51

Table 5.25: The solution of problem LA01

Job Sequence					Start Time					Finish Time				
M1	M2	M3	M4	M5	M1	M2	M3	M4	M5	M1	M2	M3	M4	M5
10	7	8	1	2	0	0	0	0	0	77	69	38	21	21
7	3	6	6	5	77	69	75	21	21	154	108	118	75	104
6	5	5	4	1	154	108	142	75	104	233	142	206	152	157
3	9	10	9	4	233	142	238	159	157	331	159	281	208	212
4	10	7	7	8	331	159	295	208	212	410	238	382	295	272
2	2	3	3	6	410	238	382	331	272	426	290	413	373	364
9	8	4	8	7	426	414	413	373	382	451	438	479	414	475
1	6	2	5	3	451	438	479	414	475	546	500	505	433	487
5	4	9	10	9	546	500	531	433	487	583	577	629	508	531
8	1	1	2	10	583	577	632	508	531	666	632	666	579	21

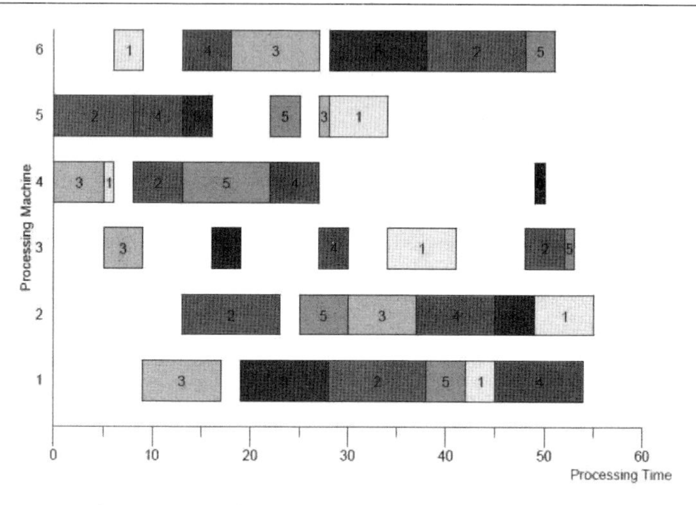

Figure 5.7 Gantt chart of the obtained solution for problem FT06.

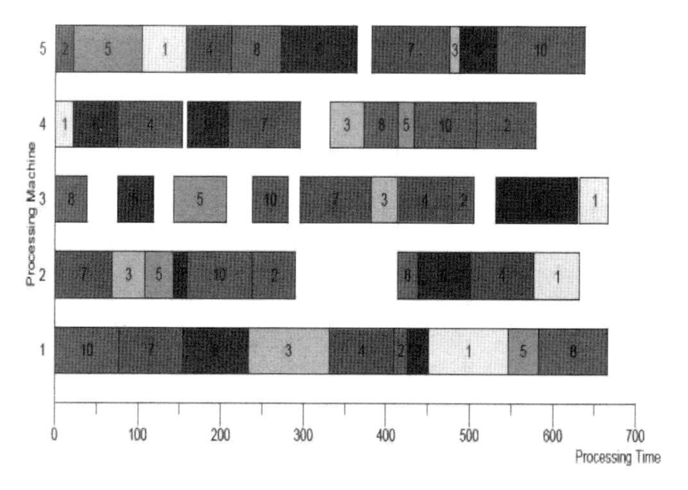

Figure 5.8 Gantt chart of the obtained solution for problem LA01.

Table 5.26: The solution of problem LA05

Job Sequence					Start Time					Finish Time				
M1	**M2**	**M3**	**M4**	**M5**	**M1**	**M2**	**M3**	**M4**	**M5**	**M1**	**M2**	**M3**	**M4**	**M5**
2	6	9	1	7	5	0	0	0	0	5	28	49	72	53
5	2	10	3	4	28	28	49	72	53	28	63	114	118	112
8	9	5	7	1	40	63	114	217	112	40	146	187	254	199
4	7	8	9	6	282	146	187	254	199	282	217	274	294	244
6	4	3	4	2	287	217	274	294	244	287	236	295	328	292
1	10	2	5	10	382	236	295	328	292	382	253	334	352	382
7	3	1	6	3	411	253	382	352	382	411	273	448	430	479
10	5	4	8	9	438	273	448	430	479	438	298	485	485	527
3	8	7	2	8	534	298	485	485	527	534	331	497	539	565
2	1	6	10	5	541	448	497	539	565	541	508	580	562	593

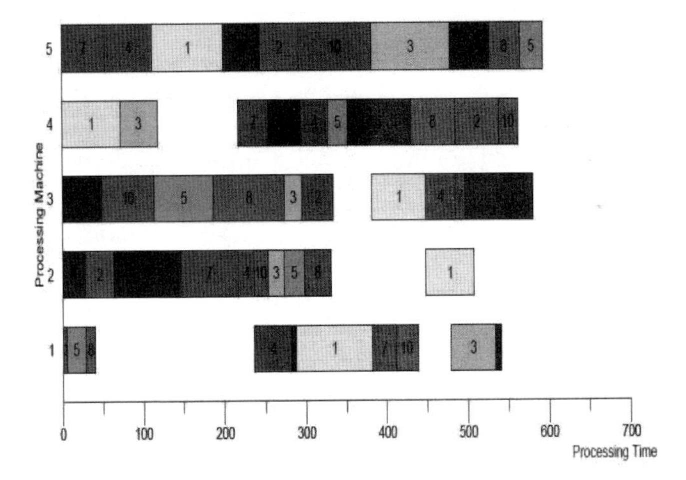

Figure 5.9 Gantt chart of the obtained solution for problem LA05.

The non-deterministic nature of our algorithm makes it necessary to carry out multiple runs on the same problem instance in order to obtain meaningful results. We ran our algorithm 15 times for each test problem.

Worst, best, mean and standard deviation (Std.) of results are illustrated in Table 5.27. The standard deviation is a very important problem in practical production area; which is calculated by:

$$std. = \sqrt{\frac{\sum_{i=1}^{i=n}(x_i - x_m)^2}{n-1}} \quad (4.1)$$

where x_i is individual result and x_m is the mean individual results.

In order to ensure our scheduling plan performs effectively, the standard deviation should be as small as possible. Through simulation, we find the standard deviation is small which indicates that HGA possesses excellent robustness.

Table 5.27: Mean, Smallest, Largest and standard deviation for each test instance FT06, (LA01-LA20).

Problem	Mean	Worst(HGLA)	Best(HGLA)	Std.(HGA)
Ft06 (6×6)	55	55	55	0
LA01 (10×5)	674.3000	680	666	5.0210
LA02 (10×5)	729	758	714	18.7457
LA03 (10×5)	653.1000	671	617	11.5884
LA04 (10×5)	630.7000	672	632	3.9000
LA05 (10×5)	593	593	593	0
LA06 (15×5)	593	593	926	0
LA07 (15×5)	934.5714	973	899	22.7564
LA08 (15×5)	863	863	863	0
LA09 (15×5)	951	951	951	0
LA10 (15×5)	958	958	958	0
LA11 (20×5)	1222	1222	1222	0
LA12 (20×5)	1039	1039	1039	0
LA13 (20×5)	1150	1150	1150	0
LA14 (20×5)	1292	1292	1292	0
LA15 (20×5)	1349.4	1379	1311	22.9371
LA16 (10×10)	1088.6	1107	1077	11.2427
LA17 (10×10)	851.929	877	843	3.9362
LA18 (10×10)	930.0714	954	894	12.8072
LA19 (10×10)	961	1015	896	42.0077
LA20 (10×10)	991	1056	949	18.3011

Our study consists of the comparison of performance with other optimization algorithms to demonstrate the efficiency and robustness of the proposed algorithm. We compare our results with results obtained by Deepanandhini and Amudha [149]. They used a nature-inspired metaheuristic algorithm called Consultant Guided Search algorithm (CGS). They solved major of our test problems [149]. Table 5.28 illustrated our results and Deepanandhini and Amudha results and the relative error for the two algorithms with respect to BKS. We note from table 5.28 our proposed algorithm (HGA) reached to the best known solutions for 52.4% of its solved test problems and the CGS reached to the best known solutions for 26.3% of its solved test problems and major remained problems HGA reached near to BKS than CGS.

Table 5.28: The comparison between the proposed algorithm result (HGA) and (CGS) results

Problem	CGS	Percentage Relative error (CGS)	HGA	Percentage Relative error (HGA)
Ft06 (6×6)	-	-	55	0.000%
LA01 (10×5)	713	7.057%	666	0.000%
LA02 (10×5)	757	15.572%	714	9.008%
LA03 (10×5)	682	14.238%	617	3.350%
LA04 (10×5)	669	13.389%	632	7.119%
LA05 (10×5)	593	0.000%	593	0.000%
LA06 (15×5)	926	0.000%	926	0.000%
LA07 (15×5)	-	-	877	1.685%
LA08 (15×5)	897	3.939%	863	0.000%
LA09 (15×5)	956	0.526%	951	0.000%
LA10 (15×5)	958	0.000%	958	0.000%
LA11 (20×5)	1222	0.000%	1222	0.000%
LA12 (20×5)	1058	1.829%	1039	0.000%
LA13 (20×5)	1152	0.174%	1150	0.000%
LA14 (20×5)	1292	0.000%	1292	0.000%
LA15 (20×5)	1310	8.536%	1311	8.616%
LA16 (10×10)	1010	6.878%	1077	13.968%
LA17 (10×10)	-	-	843	7.526%
LA18 (10×10)	914	7.783%	894	5.425%
LA19 (10×10)	944	12.114%	896	6.413%
LA20 (10×10)	949	5.211%	972	7.761%

Figure 5.10 proposed the percentage relative error for two methods (CGS) and (HGA).The lesser percentage relative error for two methods is HGA percentage

relative error and the higher is CGS values. As result, our proposed algorithm more converges to the BKS and exhibits a superior performance in comparison to CGS.

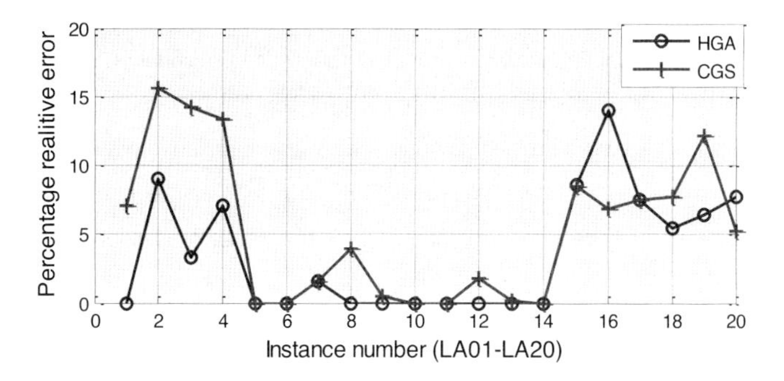

Figure 5.10 Relative Error by instance number (LA01-LA20).

5.4 Conclusion

In this chapter we presented a combination between GA and local search for solving the Job-shop scheduling problem. Firstly, a new initialization method is proposed. A modified crossover and mutation operators are used. Secondly, local search based on the neighborhood structure is applied to the GA result. Finally, the approach is tested on a set of standard instances taken from the literature. The results obtained by HGA are better than solutions obtained by GA only, which mean that hybridizing local search with GA improves the solutions quality. In addition the proposed algorithm is more converges to the BKS and gives comparable minimum relative error or better than those obtained by other approach. In general, the results have shown the correctness, feasibility and usability of the proposed algorithm

CHAPTER 6

Conclusions and Future Work

6.1 Conclusions

The main conclusions of this thesis and the results which obtained in the previous chapters can be summarized as follows:

1- We investigate new efficiency optimization method for nonlinear optimization problems by implementing a local search referred to chaos search incorporated with genetic algorithm, It can provide more efficient behavior and a higher flexibility for nonlinear optimization problems. Technique is tested by solving constrained and unconstrained test problems and the resulted outcomes are analyzed. A careful observation will reveal the following benefits of the proposed optimization algorithm:

- CGA integrates the powerful global searching capability of the GA with the powerful local searching capability of the Chaos search.
- Unlike classical techniques CGA search from a population of points, not single point. Therefore,it can provide a globally optimal solution.
- CGA uses only the objective function information, not derivatives or other auxiliary knowledge. Therefore it can deal with the non-smooth, non-continues and non-differentiable functions which are actually existed in practical optimization problems.
- CGA found the better solutions than the other methods that reported in the literature
- CGA was demonstrated to be extremely effective and efficient at locating optimal solutions.

- Due to simplicity of CGA procedures, it can be used to handle complex problems of realistic dimensions.

2- We proposed a new approach (HGA) for job shop scheduling problem. HGA is combination between GA and local search method for solving the Job-shop scheduling problem. First, GA is applied to JSSP .A new initialization method is proposed. Advanced crossover and mutation operators are used. Secondaly, local search based on the neighborhood structure is applied in the GA result. The approach is tested on a set of standard problems taken from the literature. The obtained result illustrates that the proposed algorithm is able to obtain the optimum or a useful near optimum result for the JSSP problems. Showing that using local search with GA has better performance than GA only and it is possible to accelerate convergence speed, good quality of optimal solutions and more predictable application behavior. The computation results validate the effectiveness of using genetic algorithm and local search for the job-shop scheduling problem.

6.2 Future Work

Some possible areas for further work have arisen from this research :

1) Solving larger scale nonlinear examples to demonstrate the efficiency of CGA.
2) Updating CGA to solve the multi-objective optimization problems.
3) Using other chaotic maps to accelerate the convergence property of the proposed algorithm and improve the solution quality.
4) Solving larger scale job-shop scheduling problem examples to demonstrate the efficiency of the proposed algorithm.
5) Hybridizing more optimization techniques with the proposed algorithm to accelerate the convergence property and improve the solution quality.
6) Updating the HGA to solve the JSSP as multi-objective optimization problem

Bibliography

Bibliography

[1] Forst W., Hoffmann D., (2010), "*Optimization—Theory and Practice*", Springer Science & Business Media, New York, first edition.

[2] Craven B. D., (1995), "*Control and Optimization*", Chapman and Hall, London, first edition.

[3] Rao S.S., (2009), "*Engineering Optimization: Theory and Practice*", A Wiley-Interscience publication, New Jersey, fourth edition.

[4] Snyman J. A., (2005), " *Practical Mathematical Optimization: An Introduction to Basic Optimization Theory and Classical and New Gradient-Based Algorithm* ", Springer Science & Business Media, New York.

[5] Boggs P.T., Tolle J.W., (2000), "*Sequential quadratic programming for large-scale nonlinear optimization* ", Journal of Computational and Applied Mathematics, 124(1-2), 123-137.

[6] Michalewicz Z., (1994), "*Evolutionary Computation Techniques for Nonlinear Programming* Problems ", International Transactions in Operational Research, 1(2), 223–240.

[7] Onwubolu G.C., B.V.Babu, (2004), "*New optimization Techniques in Engineering* ", Springer Science & Business Media, Berlin, first edition.

[8] Birge J. R., Louveaux F., (2011), "*Introduction to stochastic programming*", Springer Science & Business Media, New York, second edition.

[9] Stefanov S.M., (2001), "*Separable Programming: Theory and Methods*", Springer Science & Business Media, US, first edition.

[10] Deb K, (2001), "Multi-objective optimization using evolutionary algorithms", John Wiley & Sons, New York, first edition.

[11] Foulds L.R., *(1981),* "Optimization Techniques *", Springer Science & Business* Media, New York, first edition.

[12] Swann W. H., (1969), "*A survey of nonlinear optimization techniques*", Federation of European Biochemical Societies Letters, 2(1), S39-S55

Bibliography

[13] Bertsekas D.P., (1999), "*Nonlinear programming*", Athena Scientific, USA, 2nd edition.

[14] Bertsekas D.P, (1996), "*Constrained Optimization and Lagrange Multiplier Methods*", Athena Scientific, Belmont, First edition.

[15] Dussault J.P, (1995), "*Numerical stability and efficiency of penalty algorithms*", SIAM Journal on Numerical Analysis, 32(1), 296–317.

[16] Yeniay Q., (2005), "*Penalty Function For Constrained Optimization With Genetic Algorithm* ", Mathematical and Computational Applications, 10(1), 45-56.

[17] Fiacco A.V. and McCormick G.P., (1990), "*Nonlinear Programming: Sequential Unconstrained Minimization Techniques*", John Wiley & Sons, New York, First edition.

[18] Murphy F. H., (1974), "*A Class of Exponential Penalty Functions*", SIAM Journal Control, 12(4), 679-687.

[19] More J.J. and Sorensen D.C., (1983), "*Computing a trust region step*", SIAM Journal on Scientific and Statistical Computing, 4(3), 553–572.

[20] Philip E. G., Walter M., Michael A. S. and Margaret H. W., (1988), "*Recent developments in constrained optimization*", Journal of Computational and Applied Mathematics, 22(2-3), 257-270.

[21] Wilson R.B., (1962), "*A simplicial algorithm for concave programming*", PhD thesis, Graduate School of Business Administration, Harvard University.

[22] Han S.P., (1977), "*A globally convergent method for nonlinear programming*", Journal of Optimization Theory and Applications, 22(3), 297–309.

[23]Han S.P., (1976), "*Superlinearly convergent variable metric algorithms for general nonlinear programming problems*", Mathematical Programming, 11(1), 263-282.

[24] Powell M.J.D., (1978), "*Algorithms for nonlinear constraints that use Lagrangian functions*", Mathematical Programming, 14(1), 224–248.

[25] Schittkowski K., (1985), "*NLPQL: A fortran subroutine solving constrained*

nonlinear programming problems", Annals of Operations Research, 5(2), 485-500.

[26] Schittkowski K., (1983), "*On the convergence of a sequential quadratic programming method with an augmented Lagrangian line search function*", Mathmatische Operations forschung und Statistik. Series Optimization, 14(2), 197-216.

[27] Lalee M., Nocedal J., and Plantenga T., (1998), "*On the implementation of an algorithm for large-scale equality constrained optimization*", SIAM Journal on Optimization, 8(3), 682-706.

[28] Murray W. and Prieto F.J., (1995), "*A sequential quadratic programming algorithm using an incomplete solution of the subproblem*", SIAM Journal on optimization, 5(3), 590-640.

[29] Yang X., Yuan J., Yuan J., and Mao H., (2007), "*A modified particle swarm optimizer with dynamic adaptation*", Applied Mathematics and Computation, 189(2), 1205-1213.

[30] Nocedal J., Wright S., (2006), "*Numerical optimization*", Springer, New York, second edition.

[31] Haupt R. L., Haupt S.E., (2004), "*practical genetic algorithm*", A Wiley-Interscience publication, New Jersey, second edition.

[32] Brooks S.P., Morgan B.J.T, (1995), "*optimization using simulated annealing*", Journal of the Royal Statistical Society, 44(2), 241-257.

[33] Alrefaei M. H. and Diabat A.H., (2009), "*A simulated annealing technique for* multi-*objective simulation optimization*", Applied Mathematics and Computation, 215(8), 3029–3035.

[34] Hopfield J. J. and Tank D. W., (1985), " *"Neural" Computation of Decisions in Optimization Problem*", Biological Cybernetics, 52, 141-152.

[35] Liu M., Wang R., Wu J. and Kemp R., (30 May- 1 June 2005), "*A Genetic-Algorithm-Based Neural Network Approach for Short-Term Traffic Flow*

Forecasting", Advances in Neural Networks- Second International Symposium on Neural Networks, China.

[36] Glover F. and Laguna M., (1997), "*Tabu Search*", Kluwer Academic Publishers, Norwell, MA, USA, First edition.

[37] Laguna M., (1994), "*A guide to implementing Tabu Search*", Investigacion Operativa, 4(1), 5-25.

[38] Dreo J. and Siarry P., (2006), "*An ant colony algorithm aimed at dynamic continuous optimization* ", Applied Mathematics and Computation, 181(1), 457-467.

[39] Dorigo M., Stützle T., (2004), "*Ant Colony* Optimization", Bradford Company Scituate, MA, USA, First edition.

[40] Panigrahi S. K., Sahub A., Pattnaik S., (2015), "*Structure Optimization Using Adaptive Particle Swarm Optimization*", Procedia Computer Science, 48, 802-808.

[41] Khan S., Asjad M. and Ahmed A., (2015), "*Review of Modern Optimization Techniques*", International Journal of Engineering Research & Technology, 4(04), 948-988.

[42] Singh J., Chutani S., (2015), "*A Survey of Modern Optimization Techniques for Reinforced Concrete Structural Design* ", International Journal of Engineering Science Invention Research & Development, 2(1), 55-62.

[43] Abd El-Wahed W.F., Mousa A.A., El-Shorbagy M.A., (2011), "*Integrating Particle Swarm Optimization with Genetic Algorithms for Solving Nonlinear Optimization Problems*", Journal of Computational and Applied Mathematics, 235(5), 1446–1453.

[44] Gallagher K. and Sambridge M., (1994), "*Genetic algorithms: A powerful tool for large-scale nonlinear optimization problems*", Compute and Geosciences, 20(7/8), 1229-1236.

Bibliography

[45] Aggarwal S., Garg R., Goswami P., (2014), "*A Review Paper on Different Encoding Schemes used in Genetic Algorithms* ", International Journal of Advanced Research in Computer Science and Software Engineering, 4(1), 590-600.

[46] Malhotra R., Singh N., Singh Y., (2011), "*Genetic algorithms: Concepts, design for optimization of process controllers*", Computer and Information Science, 4(2), 39-54.

[47] Shukla A., Pandey H. M., Mehrotra D., (25-27 Feb. 2015) , "*Comparative Review of Selection Techniques in Genetic Algorithm*", International Conference on Futuristic Trends on Computational Analysis and Knowledge Management, Noida, 515-519.

[48] Jebari K., Madiafi M., (2013) , "*Selection Methods for Genetic Algorithms*", International Journal of Emerging Science and Engineering, 3(4), 333-344.

[49] Baker J. E., (28-31 July 1987), "*Reducing bias and inefficiency in the selection algorithm* ", Proceedings of the second international conference on genetic algorithms and their application, Hillsdale, New Jersey, USA , 14-21.

[50] Soni N., Kumar T., (2014), "*Study of Various Crossover Operators in Genetic Algorithms*", International Journal of Computer Science and Information Technologies, 5 (6), 7235-7238.

[51] *Soni N., Kumar T., (2014), "Study of Various Mutation Operators in Genetic Algorithms",* International Journal of Computer Science and Information Technologies, 5 (3), 4519-4521

[52] Villagra A., Pandol D., Rasjido J., Montenegro C., Seron N., Leguizamon G. and Guillermo M., (18-22 Dec. 2010), "*Repair Algorithms and Penalty Functions to Handling Constraints in an Evolutionary* Scheduling", XVI Argentine Congress of Computer Science, University of Morón, Argentine.

[53] Kuma A., (2011), "*Network Design Using Genetic Algorithm* ", A Thesis submitted to saurshtra university, Rajkot, India.

[54] Biggs M. B., (2008), "*Nonlinear Optimization with Engineering Applications* ", Springer, Verlag , first edition.

[55] Azzam M. and Mousa A. A., (2010), "*Using genetic algorithm and TOPSIS technique for multiobjective reactive power compensation*", Electric Power Systems Research, 80(6), 675–681.

[56] Mousa A.A., (2010), "*Using genetic algorithm and TOPSIS technique for multiobjective transportation problem: a hybrid approach*", International Journal of Computer Mathematics, 87(13), 3017-3029.

[57] Mousa A.A., El-Shorbagy M.A. and Abd-El-Wahed W.F., (2012) ,"*Local search based hybrid particle swarm optimization algorithm for multiobjective optimization*", Swarm and Evolutionary Computation, 3, 1–14.

[58] Mousa A.A., Abd El-Wahed Waiel F., Rizk-Allah R. M., (2011), "*A hybrid ant colony optimization approach based local search scheme for multiobjective design optimizations*", Electric Power Systems Research, 81, 1014–1023.

[59] Mousa A.A., (2014), "*Hybrid Ant Optimization System for Multiobjective Economic Emission Load Dispatch Problem Under Fuzziness*", Swarm and Evolutionary Computation, 81(4), 11–21.

[60] Osman M.S., Abo-Sinna M.A., and Mousa A.A., (2005), "*A Combined Genetic Algorithm-Fuzzy Logic Controller (GA-FLC) In Nonlinear Programming*", Journal Of Applied Mathematics & Computation (AMC), 170(2), 821-840.

[61] Fares M., Kaminska B., (1995), "*A Fuzzy Nonlinear Programming Approach to* Analog Circuit Design", *IEEE Transactions on Computer-Aided Design of Integrated Circuits and Systems,* 14(17), 785-793.

[62] Elsayed S. M., Sarker R. A. and Essam D. L., (2014), "*A new genetic algorithm for solving optimization problems*", Engineering Applications of Artificial Intelligence, 27, 57–69.

[63] Deb K., (1991), "Optimal design of a welded beam via genetic algorithms", The American Institute of Aeronautics and Astronautics Journal, 29(11), 2013-2015 .

[64] Hussein M.A., EL-Sawy A.A., Zaki E.M., and Mousa A.A., (2014), "*Genetic algorithm and rough sets based hybrid approach for economic environmental dispatch of power systems*", British Journal of Mathematics & Computer Science, 4(20), 2978-2999.

[65] Tsoulos I. G., (2009), "*Solving constrained optimization problems using a novel genetic algorithm*", Applied Mathematics and Computation, 208(1), 273-283.

[66] Juan W., Ping W., (17-19 Dec 2010), "*Optimization of Fuzzy Rule Based on Adaptive Genetic Algorithm and Ant Colony Algorithm*", International Conference on Computational and Information Sciences (ICCIS), Chengdu, 359-362.

[67] Sun F., Tian Y., (17-19 Dec. 2010),"*Transmission Line Image Segmentation Based GA and PSO Hybrid Algorithm, Computational and Information Sciences*" , International Conference on Computational and Information Sciences (ICCIS), Chengdu, 677-680 .

[68] Donis-Díaz C.A., Muro A.G. and Bello-Pérez R. and Morales E.V., (2014), "*A hybrid model of genetic algorithm with local search to discover linguistic data summaries from creep data*", Expert Systems with Applications, 41(2), 2035–2042.

[69] Sawyerr B.A., Adewumi A.O. and Ali M.M., (2014), "*Real-coded genetic algorithm with uniform random local search*", Applied Mathematics and Computation, 228, 589–597.

[70] Yang K., Liu Y. and Yang G., (2013) "*Solving fuzzy p-hub center problem by genetic algorithm incorporating local search*", Applied Soft Computing, 13(5), 2624–2632

[71] Derbel H., Jarboui B., Hanafi S. and Chabchoub H., (2012), "*Genetic algorithm with iterated local search for solving a location-routing problem*", Expert Systems with Applications, 39(3), 2865–2871.

[72] Kabir Md. M., Shahjahan Md. and Murase K., (2011), "*A new local search based hybrid genetic algorithm for feature selection*", Neurocomputing, 74(17), 2914–2928.

[73] Kilani Y., (2010), "*Comparing the performance of the genetic and local search algorithms for solving the satisfiability problems*", Applied Soft Computing, 10(1), 198–207.

[74] Hénon M., (1976), "*A two-dimensional mapping with a strange attractor*", Communications in Mathematical Physics, 50(1), 69-77.

[75] Lorenz E.N., (1993), "*The Essence of Chaos*", University of Washington Press, USA, first edition

[76] Liu B., Wang L., Jin Y., Tang F. and Huang D., (2005), "*Improved particle swarm optimization combined with chaos*", Chaos, Solitons and Fractals, 25(5), 1261–1271.

[77] Xiang T., Liao X., Wong K., (2007), "*An improved particle swarm optimization algorithm combined with piecewise linear chaotic map*", Applied Mathematics and Computation, 190(2), 1637–1645.

[78] Wang L., Zheng D.Z., Lin Q.S., (2001), "*Survey on chaotic optimization methods*", Computing Technology and Automation, 20(1), 1–5.

[79] Coellho L. d., Mariani V.C., (2009), "*A novel chaotic particle swarm optimization approach using Hénon map and implicit filtering local search for economic load dispatch*" Chaos, Solitons and Fractals, 39(2), 510–518.

[80] Chuanwen J., Bompard E., (2005), "*A self-adaptive chaotic particle swarm algorithm for short term hydroelectric system scheduling in deregulated environment*", Energy Conversion and Management, 46(17), 2689–2696.

[81] Caponetto R., Fortuna L., Fazzino S., Xibilia M., (2003), "*Chaotic sequences to improve the performance of evolutionary algorithms*", IEEE Transactions on Evolutionary Computation, 7(3), 289 - 304.

Bibliography

[82] Tavazoei M. S., Haeri M., (2007), "*An optimization algorithm based on chaotic behavior and fractal nature*", Journal of Computational and Applied Mathematics, 206(2), 1070–1081.

[83] Yang D., Li G., Cheng G., (2007), "*On the efficiency of chaos optimization algorithms for global optimization*", Chaos, Solitons and Fractals, 34(4), 1366–1375.

[84] Cong S., Li G., Feng X., (9-11 June 2010), "*An improved algorithm of chaos optimization*", 8th IEEE International Conference on Control and Automation (ICCA), Xiamen, 1196 – 1200.

[85] Hu. W., Liang H., Peng C., Du. B. and Hu. Q., (2013), "*Hybrid Chaos-Particle Swarm Optimization Algorithm for the Vehicle Routing Problem with Time Window*", Entropy, 15(4), 1247-1270.

[86] Ebrahimzadeh R., Jampour M., (2013), "*Chaotic Genetic Algorithm based on Lorenz Chaotic System for Optimization Problems*", Intelligent Systems and Applications, 5(5), 19-24.

[87] Xiao J., (2009) "*Improved Quantum Evolutionary Algorithm Combined with Chaos and Its* Application", Springer, Berlin, Heidelberg, First edition.

[88] Alatas B., Akin E., Ozer A. B., (2009). "*Chaos embedded particle swarm optimization algorithms*", Chaos, Solitons and Fractals, 40(4), 1715-1734 .

[89] Zelinka I., Celikovsky S., Richter H. and Chen G., (2010), "*Evolutionary Algorithms and Chaotic Systems*", Springer, Berlin, Heidelberg, first edition.

[90] Coelho L., Ayal1a H., Mariani V. (2014), "*A self-adaptive chaotic differential evolution algorithm using gamma distribution for unconstrained global optimization*", Applied Mathematics and Computation, 234, 452-459.

[91] Lu P., Zhou J., Zhang H., Zhang R. and Wang C., (2014), "*Chaotic differential bee colony optimization algorithm for dynamic economic dispatch problem with valve-point effects*", International Journal of Electrical Power & Energy Systems, 62, 130–143.

[92] Tavazoei M.S., Haeri M., (2007) , "*Comparison of different one-dimensional maps as chaotic search pattern in chaos optimization algorithms*", Applied Mathematics and Computation , 187(2), 1076-1085.

[93] Hilborn R., (2004), "*Chaos and Nonlinear Dynamics: An Introduction for Scientists and Engineers*", Oxford University Press, New York, second edition.

[94] He D., He C., Jiang L. and Zhu H., (2001), "*Chaotic characteristic of a one dimensional iterative map with infinite collapses*", IEEE Transactions Circuits Syst., 48 (7), 900–906.

[95] Chen G., Huang Y., (2011), " *Chaotic Maps: Dynamics, fractals and Rapid Fluctuations* ", Morgan Kaufmann, San Mateo, California, First edition.

[96] May R. M., (1976), "*Simple mathematical models with very complicated dynamics*", Nature, 261(4), 59–67.

[97] Arora J.S., Elwakeil O.A., Chahande A.I., Hsieh C.C., (1995), "*Global optimization methods for engineering application: a review*", Structural Optimization, 9(3-4), 137–159.

[98] Li Y., Deng S., Xiao D., (2011), "*A novel Hash algorithm construction based on chaotic neural network*", Neural Computing and applications, 20(1), 133–141.

[99] Devaney R.L., (2003), "*An Introduction to Chaotic Dynamical Systems*", Westviev press, USA, second edition.

[100] Peitgen H., Jurgens H., Saupe D., (2004), "*Chaos and Fractals*", Spring Science & Business Media, New York, second edition.

[101] El-Shorbagy M.A., Mousa A.A., Abd-El-Wahed W.F., (2011), "*Hybrid Particle Swarm Optimization Algorithm for Multi-Objective Optimization*", Lambert academic publishing GmbH &Co.kG, Berlin.

[102] Popov A., (2003), "*Genetic algorithm for optimization – Application in Controller Design Problems* ", A thesis submitted to Technical University of Sofia, Bulgaria.

Bibliography

[103] Neves N., Nguyen A. and Torre E. L., (12-16 August 1996), "*A study of nonlinear optimization problems using a distributed genetic algorithm*", 25th International Conference on Parallel Processing, Ithaca, New York, 29-36.

[104] A. Auger, Hansen N., (2-5 Sept. 2005), "*A restart CMA evolution strategy with increasing population size*", Proceedings of the 2005 IEEE Congress on Evolutionary Computation, Edinburgh, United Kingdom, 1769–1776.

[105] Ammaruekarat P., Meedsad P., (28-29 May 2011), "*A Chaos Search for Multi-Objective Memetic Algorithm*", International Conference on Information and Electronics Engineering, Bangkok, Thailand, 140-144.

[106] Suganthan P. N., Hansen N., Liang J.J., Deb K., Chen Y.-P., Auger and Tiwari S., (2005), "*Problem definitions and evaluation criteria for the CEC'2005 special sessionon real parameter optimization*", Nanyang Technological University, Tech. Rep., Available in http://www.ntu.edu.sg/home/epnsugan/index_files/cec-05/Tech-Report-May-30-05.pdf

[107] Sedlaczek K., Eberhard P., (30 May - 03 June 2005), "*Constrained Particle Swarm Optimization of Mechanical Systems*", 6th World Congresses of Structural and Multidisciplinary Optimization, Brazil.

[108] Lozano J. A., Larrañaga P., Inza I,, Bengoetxea E. , (2006), "*Towards a New Evolutionary Computation*", Springer, Berlin, Heidelberg, First edition.

[109] Suman S. K., Giri V. K., (2015), "*Genetic Algorithms Basic Concepts and Real World Applications*", International Journal of Electrical, Electronics and Computer Systems, 3(12), 116-123.

[110] Stephens C. R., Toussaint M., Whitley D. and Stadler P.F., (1993), " *Foundations of Genetic Algorithms* ", Springer Science & Business Media, Berlin, First edition.

[111] Fernandes C., Rosa A., (27-30May 2001), " *A study of non-random matching and varying population size in genetic algorithm using a royal road function*", in:

Proceedings of the 2001 Congress on Evolutionary Computation, Piscataway, New Jersey, 60–66.

[112] Mülenbein H., Schlierkamp-Voosen D. (1993), "*Predictive models for the breeding genetic algorithm in continuous parameter optimization*", Evolutionary Computation, 1, 25–49.

[113] Herrera F., Lozano M. and Molina D., (2006), "*Continuous scatter search: An analysis of the integration of some combination methods and improvement strategies*", European Journal of Operational Research, 169(2), 450–476.

[114] Laguna M., Marti R., (2003), "*Scatter Search: Methodology and Implementation in C*", Springer Science & Business Media, New York, First edition.

[115] Pinedo M. L., (2009), "*Planning and scheduling in manufacturing and services*" Springer Science & Business Media, New York, second edition.

[116] Qin A. K., Suganthan P.N., (2-5 Sept. 2005), "*Self-adaptive differential evolution algorithm for numerical optimization*", in: Proceedings of the 2005 IEEE Congress on Evolutionary Computation, University of Edinburgh, United Kingdom, 1785–1791.

[117] Kolharkar S., Zanwar R., (2013), "*Scheduling in Job Shop Process Industry*", Journal of Mechanical and Civil Engineering, 5(1), 2278-1684.

[118] Chryssolouris G., Subramaniam V., (2001), "*Dynamic scheduling of manufacturing job shops using genetic algorithms*", Journal of Intelligent Manufacturing , 12(3), 281-293.

[119] Arisha A., Young P., El Baradie M., (1-3 July 2001), "*Job Shop Scheduling Problem: an Overview*", International Conference for Flexible Automation and Intelligent Manufacturing (FAIM 01), Dublin, Ireland, 682 – 693.

[120] Garey, M.R., Johnson, D.S., and Sethi, (1976), "*The Complexity of Flow-shop and Job-shop Scheduling*", Mathematics of Operations Research, 1(2), 117-129.

Bibliography

[121] Garey M.R., Johnson D.S., (1979), "*Computers and intractability: A guide to the theory of NP-completeness*", W. H. Freeman & Co., New York, first edition.

[122] Qing-dao-er-ji R., Wang Y., (2012), "*A new hybrid genetic algorithm for job shop scheduling problem*", Computers & Operations Research, 39(10), 2291–2299.

[123] Vilcaut G., Billaut J., (2008), "*A tabu search and a genetic algorithm for solving a bicriteria general job shop scheduling problems*", European Journal of Operational Research, 190, 398–411.

[124] Ghiani G., Grieco A., Guerrieri A., Manni A. and Manni.E., (2015), "*A fast heuristic for large-scale assembly job shop scheduling problems with bill of materials*", Advances in Mathematics and Statistical Sciences, 40, 216–223.

[125] Gromicho J., Hoorn J. , Saldanhada-Gama J. and Timmer G., (2012), "*Solving the job shop scheduling problem optimally by dynamic programming*", Computers & Operations Research, 39(12), 2968–2977.

[126] Moin N., Sin O. and Omar M., (2015), " *Hybrid Genetic Algorithm with Multiparents Crossover for Job Shop Scheduling Problems*", Mathematical Problems in Engineering, 2015, 1-12.

[127] Chakraborty S. and Bhowmik S., (2015), "*An Efficient Approach to Job Shop Scheduling Problem using Simulated Annealing*", International Journal of Hybrid Information Technology, 8(11), 273-284.

[128] Yousefi M., Yousefi M., Hooshyar D. and Oliveira J., (2015), "*An evolutionary approach for solving the job shop scheduling problem in a service industry*", international journal of advances in intelligent informatics,1(1), 1-6.

[129] Ponsich A., Coello C., (2013), "*A hybrid Differential Evolution—Tabu Search algorithm for the solution of Job-Shop Scheduling Problems*", Applied Soft Computing, 13, 462-474

[130] Zhang R., Chang P., Wu C., (2013), "*A hybrid artificial bee colony algorithm for the job shop scheduling problem*", Int. J. Production Economics, 141, 167-178.

Bibliography

[131] Mousa A.A., El-Sawy A.A., Hendawy Z.M., El-Shorbagy M.A., (7-9 Dec. 2012), "*Trust-Region Algorithm Based Local Search for Multi-objective Optimization*", Proceedings of the first International Conference on Innovative Engineering Systems (ICIES), Egypt, 207-212.

[132] Sreenivas P., Kumar S. V., (2015), "*A Review on Non-Traditional Optimization Algorithm for Simultaneous Scheduling Problems*", Journal of Mechanical and Civil Engineering, 12(2), 50-53.

[133] Blazewicz J., Docmschke W. and pesch E., (1996), "*The job shop scheduling problem: Conventional and new solution techniques. European Journal of Operational* ", European Journal of Operational Research , 93(1),1-33.

[134] Jones A., (1998), "*Survey of Job Shop Scheduling Techniques*", A Thesis submitted to National Institute of Standards and Technology, a unit of the U.S. Commerce Department.

[135] Balas E., (1969), " *Machine sequencing via disjunctive graphs: an implicit enumeration algorithm*", Operations Research, 17(6), 941–957.

[136] Mallikarjuna K., Venkatesh G. and Somanath B., (2014), " *A Review On Job Shop Scheduling Using Non-Conventional Optimization Algorithm* ", International Journal of Engineering Research and Applications, 4(3), 11-19

[137] Jain A. S., Meeran S., (1998), "*A state of the art review of job shop scheduling techniques*", Technical report, Department of Applied Physics, Electronic and Mechanical Engineering, University of Dundee, Dundee, Scotland.

[138] Sahu L. K., Sridhar K., (2015), "*Shiftting Bottleneck Algorithm for Job Shop Scheduling*", International Journal of Scientific Engineering and Applied Science, 2(13), 215-219.

[139] Jaziri W., (2008), " *Local Search Techniques: Focus on Tabu Search*", I-Tech Education and Publishing, Austria, First published.

[140] Zhang C., Li P., Guan Z. and Rao Y.,(2007), "*A tabu search algorithm with a new neighborhood structure for the job shop scheduling problem*", Computers & Operations Research, 34(11), 3229–3242.

[141] Zhang C., Rao Y. and Li P., (2008), "*An effective hybrid genetic algorithm for the job shop scheduling problem*", The International Journal of Advanced Manufacturing Technology, 39, 965–974.

[142] Yan-Fang Y., Yue Y., (2015), "*An improved genetic algorithm to the job shop scheduling problem*", Journal of Chemical and Pharmaceutical Research, 7(4), 322-325.

[143] Nazif H., (2015), "*Solving job shop scheduling problem using an ant colony algorithm*", Journal of Asian Scientific Research, 5(5), 261-268.

[144] Flórez E., Gómez W., Bautista L., (2013), "*An ant colony optimization algorithm for job shop scheduling problem*", International Journal of Artificial Intelligence & Applications, 4(4), 53-66.

[145] Thamilselvan R., Balasubramanie P., (2012), "*Integrating Genetic Algorithm, Tabu Search and Simulated Annealing For Job Shop Scheduling Problem*", International Journal of Computer Applications, 48(5), 42-54.

[146] Song X., (18-20 Oct. 2008), "*Deadlocks Solving Strategies in Hybrid PSO Algorithm for JSSP* ", 2008 Fourth International Conference on Natural Computation, Jinan, 614-618.

[147] Sha D.Y., Hsu C.-Y., (2006), "*A hybrid particle swarm optimization for job shop scheduling problem*", Computers & Industrial Engineering, 51(4),791–808.

[148] Zhang G., Shao X., Li P. and Gao L., (2009), "*An effective hybrid particle swarm optimization algorithm for multi-objective flexible job-shop scheduling problem*", Computers & Industrial Engneering, 56(4), 1309–1318.

[149] Deepanandhini D., Amudha T., (2013), "*Solving job-shop scheduling problem with consultant guided search metaheuristics*", International Journal of Software and Web Sciences, 3(1), 1-6.

Bibliography

[150] Lazár I., Malindžák S., (2012), " *review on solving the job shop scheduling problem techniques: recent development and trends*", Transfer inovácií, 23, 55-60.

[151] Park B. J., Choi H.R., Kim H.S., (2003), " *A hybrid genetic algorithm for the job shop scheduling problem*", Computers & Industrial Engineering, 45(4), 597–613.

[152] LI Y., CHEN Y., (2010), "*A Genetic Algorithm for Job-Shop Scheduling*", journal of software, 5(3), 269-247.

[153] Tamilarasi A., Jayasankari S., (2012), " *Evaluation on GA based Model for solving JSSP*", International Journal of Computer Applications, 43(7), 975 – 8887.

[154] Sureshkumar S., Thiruvenkadam S., (2015), "A New Combined Approach for Optimizing Makespan in Job Shop Scheduling Problem", International Journal of Innovative Research in Science, Engineering and Technology, 4(12), 11825-11829.

[155] Spanos A.C., Ponis S.T., Tatsiopoulos I. P., Christou I.T. and Rokou E., (2014), "*A new hybrid parallel genetic algorithm for the job-shop scheduling problem*", International transactions in operational research, 21(3), 479–499.

[156] Camino R.V., Varela R. , González M.A., (2010), "*Local search and genetic algorithm for the job shop scheduling problem with sequence dependent setup times*", Journal of Heuristics,16(2), 139-165.

[156] Ombuki B.M., Ventresca M., (2004), "*Local Search Genetic Algorithms for the Job Shop Scheduling Problem*", Applied Intelligence, 21(1), 99-109.

[157] Wang X., Duan H., (2014), "*A hybrid biogeography-based optimization algorithm for job shop scheduling problem*", Computers & Industrial Engineering, 73, 96–114.

[158] Beasley E.J., (1990), "*OR-library: distributing test problems by electronic mail*", Operational Research Society 41 (11), 1069-1072, http://mscmga.ms.ic.ac.uk.

Bibliography

[159] Gen M., Cheng R., (1997), "*Genetic algorithms and engineering design*", John Wiley-Interscience Publication, Canda.

[160] Goncalves J.F., Mendes J.J.M., Resende M.G.C., (2005), "*A hybrid genetic algorithm for the job shops scheduling problem*", European Journal of Operational Research, 167(1), 77–95.

[161] El-Shorbagy M.A., Mousa A.A., Nasr S.M., (2016), "*A Chaos-Based Evolutionary Algorithm for General Nonlinear Programming Problems*", Chaos, Solitons and Fractals, 85, 8–21.

[162] Starkweather T., Whitley D., Mathias K., Mcdaniel S., (1992)," *New Directions for Operations Research in Manufacturing* ", Springer, Berlin Heidelberg, First edition.

[163] Nasr S.M., El-Shorbagy M. A., El-Desoky I.M., Hendawy Z.M., Mousa A.A., (2015), "*Hybrid genetic algorithm for constrained nonlinear optimization problems*", British journal of mathematics & computer science, 7(6), 466-480.

[164] Cheng R., Gen M., Tsujimura Y., (1999), "*A tutorial survey of job-shop scheduling problems using genetic algorithms*", Computers & Industrial Engineering, 36(2), 343-364.

[165] Gao J., Sun L., Gen M., (2008), "*A hybrid genetic and variable neighborhood descent algorithm for flexible job shop scheduling problems*", Computers & Operations Research, 35(9), 2892-2907.

◎编辑手记

世界著名数学家 C. U. P. M. Panel 曾指出：

> 广博，包括应用领域的知识，对最有意义的数学研究具有重大价值. 对于那些有能力学习各种课程的学生，重要的是把他所能吸收的知识全部用来构成关于数学的完整图画.

本书就是一部很好的践行这一论断的英文数学专著，是我们工作室从国外引进的.

本书的中文书名或可译为《一种基于混沌的非线性最优化问题：作业调度问题》.

本书共有三位作者：M. A. 艾尔－萨尔巴吉(M. A. El-Shorbagy)，埃及数学家，姆努菲亚大学工学院数学教授.

S. 纳斯尔(S. Nasr),埃及数学家,姆努菲亚大学工学院教授.

阿卜杜拉·A. 穆萨(Abd allah A. Mousa),埃及数学家,姆努菲亚大学工学院数学教授,发表过50余篇论文,指导过10余篇论文.

正如本书作者所指出的那样:

> 最优化问题是许多科学和工程学科的主要研究课题,该领域目前仍然存在很多开放性问题.生产调度是生产组织的物流绩效的一个核心要素.生产调度最优化的重要性日益增长,即大部分公司基于自身解决实际问题的需要,发展其生产管理方式.作业调度问题(JSSP)是生产调度的一个分支,也是最难的组合优化问题之一.
>
> 为了解决最优化问题,出现了多种技术.遗传算法是最优化界中日益备受关注的一种优化技术.混沌是普遍存在于科学各个领域中的非线性现象.近些年来,混沌理论已经应用到最优化科学中的许多方面.
>
> 在本书中,我们展示了一种新的混合优化算法来解决最重要的最优化问题之一(非线性最优化问题).混合算法是一种遗传算法和混沌理论相结合的组合算法.将遗传算法和混沌局部搜索程序相结合,既可以展现两种最优化方法的优势,又能提高收敛性以达到全局解.我们的方法是在一系列取自文献的标准实例上进行测试的.
>
> 此外,混合遗传算法是用来解决真实的最优化问题之一(作业调度问题)的.我们的算法是遗传算法和局部

搜索的结合. 我们利用遗传算法设计了一个世代交替模型. 然后,在遗传算法结果中应用基于邻域结构的局部搜索. 本书提出的求解作业调度问题的方法在一系列取自文献的标准实例上进行了测试. 计算结果证实了该算法的有效性.

本书包含六章:

第一章是关于最优化主题的一个综述,我们提出了最优化问题的数学模型,然后介绍最优化问题的分类. 本章中我们也提出了解最优化问题的最优化方法.

第二章致力于介绍遗传算法的工作原理,并解释遗传算法是如何应用到解最优化问题中的. 我们还介绍了遗传算法参数和使用遗传算法解最优化问题的优势和劣势.

第三章提出解非线性最优化问题的一个新算法. 该算法是最优化方法之一(遗传算法) 与混沌理论的结合,以提高性能并获得最佳解决方案. 它是一个为解决非线性最优化问题而结合了遗传搜索特征和混沌搜索特征的新算法. 为了说明找到最优解的成功结果,已经测试了各种基准问题.

第四章提出了作业调度问题的结构,然后介绍了作业调度问题的公式化. 我们也展示了作业调度问题的复杂性. 最后,我们介绍了求解作业调度的方法.

第五章旨在实现我们解作业调度问题的新方法,并解释它的细节,各种基准问题的实验结果的测试也会被讨论. 我们通过将结果与另一种方法进行比较,来表明我们的方法的可靠性及其解决作业调度问题的能力.

第六章为结论,以及给未来研究者的几点建议.

本书的版权编辑李丹为我们翻译了本书的中文目录,如下:

关于本书的研究，国内也有一些，如暨南大学温录亮硕士2009年在其导师樊锁海教授的指导下完成了题为《多目标最优化方法与应用》的硕士学位论文. 在论文中他介绍道：

在经济规划、计划管理、金融决策、工程设计、城市与工农业规划、卫生保健和军事科学等社会活动中人们更多遇到的是同时追求多个目标的最优化问题，而不是单一目标的最优化问题. 例如，在商品生产规划中，往往要考虑"成本""质量"和"利润"等评价准则，并依据这些准则确定“成本最少”“质量最好”和“利润最大”等目标.如果一个方案可以满足所有的目标,这当然是理想的状态.然而这种“理想状态”一般来说是不可能达到的,或者说有些目标之间经常具有冲突性,当一个目标改善的同时又导致另一些目标的降低.这正是由于这种多目标间的冲突性才使得人们要去研究多目标问题决策理论和方法.同时要求多个目标都要尽可能好的最优化问题,称为多目标最优化问题.

1. 多目标最优化问题的发展简史

多目标最优化的思想萌芽起源于1776年经济学中有关效用理论的研究.1896年,经济学家V. Pareto首先在经济平衡中提出了多目标规划问题,引进所谓Pareto最优解的概念.此外1927年数学家F. Hausdorff对关于有序空间理论的研究,为多目标规划的发展提供了理论工具.1947年,J. von Neumann和O. Morgenstern在关于对策论的著作中提到了多目标决策问题,引起了人们对多目标优化的重视.1951年,H. W. Kuhn和A. W. Tucker关于向量极值问题的工作也为这一学科的发展奠定了良好的基础.1968年,Z. Johnsen系统地提出了关于多目标决策模型的研究报告,这是多目标优化这门学科走向迅速发展的一个转折点.20世纪70年代,多目标最优化问题越来越引起各方面的重视,陆续提出了一些求解多目标最优化的方法.1975年M. Zeleny初步做了文献总结.20世纪80年代末期,国际上对多目标进化算法MOEA(Multi-Objective Evolutionary Algorithm)的研究进入了兴旺时期，从1994年到2001年的8年中，国际上所出版的论文是过去10年(1984—1993)的3倍之多.最近几年的发展速度比过去8年又有提高，一方面，从2001年以来，每两年召开一次有关多目标进化的国际会议EMO(Evolutionary Multi-Criterion Optimization)；另一方面，从国际刊物 *IEEE Transactions on Evolutionary Computation*(1997年创刊)和 *Genetic Programming and Evolvable Machines*(1999年创刊)，以及各类的国际进化计算的会议上所发表的有关多目标进化的论文比过去有较大幅度的增加，这标志着MOEA的研究进入了快速发展阶段.特别是近10年来，理论探索不断深入，应用范围日益广泛，研究队伍迅速壮大，显示出勃勃生机.有关多目标优化的国际学术会议多次召开，这方面的论文不断发表，多目标优化正作为一个重要的数学分支被进行系统地研究.

2. 多目标最优化问题的数学模型

多目标优化问题也称多目标数学规划问题，可用如下方式表述

$$\begin{cases}\min\limits_{x\in G} F(\boldsymbol{x}) \\ G=\{\boldsymbol{x}\mid G(\boldsymbol{x})\geqslant \boldsymbol{0}\}\end{cases} \tag{1}$$

其中：$\boldsymbol{x}$ 是 n 维向量，$F(\boldsymbol{x})$ 和 $G(\boldsymbol{x})$ 分别是 $\boldsymbol{x}$ 的 m 维和 k 维向量函数，即

$$\boldsymbol{x}=(x_1,x_2,\cdots,x_n)^{\mathrm{T}}$$

$$F(\boldsymbol{x})=(f_1(x),f_2(x),\cdots,f_m(x))^{\mathrm{T}}$$

$$G(\boldsymbol{x})=(g_1(x),g_2(x),\cdots,g_k(x))^{\mathrm{T}}$$

$$(g_i(x) \geqslant 0, i=1,2,\cdots,k)$$

对于问题(1) 可给出各种意义的解. 其中关于(1) 的有效解和弱有效解的定义如下:

对任意 $i \in M=\{1,2,\cdots,m\}$,设 $\boldsymbol{x} \in G$,若有 $\{\boldsymbol{x} \mid F(\boldsymbol{x}) \leqslant F(\bar{\boldsymbol{x}}), \boldsymbol{x} \in G\}=\varnothing$,则称 $\bar{\boldsymbol{x}}$ 是(1) 的有效解.

对任意 $i \in M=\{1,2,\cdots,m\}$,设 $\boldsymbol{x} \in G$,若有 $\{\boldsymbol{x} \mid F(\boldsymbol{x}) < F(\bar{\boldsymbol{x}}), \boldsymbol{x} \in G\}=\varnothing$,则称 $\bar{\boldsymbol{x}}$ 是(1) 的弱有效解.

3. 国内外研究现状

多目标最优化问题的研究方向可以归纳为四类:(1) 解的概念及其性质的研究. 1951 年,H. W. Kuhn 和 A. W. Tucker 首先引进了一种特殊的有效解,称为 KT — 有效解. 1968 年,A. G. Geoffrion 对有效解进一步加以限制,界定一种新的解,这种新的解叫作 G— 有效解. 1973 年,S. Smale 用微分拓扑的观点研究了有效解. 1977 年,J. Borwein 借助于切锥概念引进一种真有效解,被称为 B— 真有效解. 1979 年,H. P. Benson 借助投影锥也引进一种真有效解,称为 P— 真有效解. 陈光亚在一般向量空间中研究了有效点的几何性质. 1982 年,M. T. Henigl 利用闭凸锯又引进一种真有效解,称为 H — 真有效解等. (2) 关于多目标优化的解法,这类问题又可分为传统解法和进化解法两种,后文将进行详细介绍. (3) 对偶问题的研究,可以将对偶问题分为两类:Lagrange 对偶和共轭对偶,其中日本学者 Tanino 以及 E. E. Rosinger 和 D. T. Lue 的工作最为突出. (4) 不可微多目标优化的研究. 近年来国内外学者都非常重视不可微多目标优化的研究,F. H.

Clarke,A. D. Loffe 和 J. P. Aubin 从不同的角度出发,对非光滑函数的广义梯度进行研究,中国学者史树中、陈光亚、汪寿阳和董加礼等从不同侧面对不可微优化进行了大量的研究,并取得了很多重要的成果.

4. 多目标进化算法

进化算法的研究起源于20世纪50年代末,成熟于20世纪80年代.进化算法具有四大分支:在20世纪60年代中期,Holland 提出了遗传算法(Genetics Algorithm,GA),Rosenberg 提出了进化策略(Evolutionary Strategy),Foguel 等人提出了进化规划,20世纪90年代,在遗传算法的基础上,Koza 将 GA 应用于计算机程序及自动化生成,提出了遗传程序设计,虽然这几个分支在算法实现方面有一些细微的差别和各自的特征,但它们具有一个共同的特点,即都是借鉴生物进化的思想和机理解决实际问题的.目前,进化算法已在自动控制、经济预测、机器学习、工程优化等领域得到了成功的应用,且在众多领域掀起了进化计算的研究高潮.现实世界中的很多优化问题都涉及多个目标的优化,多目标优化问题和单目标优化问题不同,单目标优化问题的最优解往往只有一个或少数几个,而多目标优化问题的最优解往往很多,甚至是无穷多个,而进化算法是以种群进化为基础的,其进化结果是一群解,利用它可在一次运行中求出问题的多个解甚至全部解,因此,进化算法模型比较适合于多目标优化问题的求解,同时,进化算法因其具有的隐并行性、智能性及表现的自适应性和自组织性等特点,加之其又适应于大规模的并行运算，可以搜索到问题空间中的多个区域.因此，进化算法为多目标优化问题的解决提供了新的思路和新的契机.

另外，由于多目标优化问题的广泛存在性和求解的困难性，该问题一直是富有挑战性和吸引力的研究课题. 20世纪90年代开始流行的进化算法为求解多目标优化问题提供了有力的工具.近年来，进化计算界相继提出了不同的多目标进化算法，它们的提出引起了众多研究机构的重视，这一方向已成为学术界研究的热点.近几年来，这一方面取得的令人瞩目的成就与其应用于工程中的巨大价值已取得IEEE学会的《进化计算杂志》为其专门出版的多目标进化计算理论与技术的专辑.

(1) 多目标进化算法的分类

自1985年Schaffer提出第一个多目标进化算法以来，研究者采用各自不同的适应值分配策略、选择策略、多样性保持策略、精英策略、约束条件处理策略等对简单进化算法进行修改，也因此涌现出众多优秀的多目标进化算法，形成了一个热门的研究领域.而我们根据决策者对各目标主观偏重程度与搜索过程的相互影响关系，多目标进化算法可以分为先验方式、交互式和后验方式.

先验方式是在算法进行搜索前，决策者事先设置各目标的优先权值，将全体目标按权值合成一个标量目标函数，将多目标优化问题转化为单目标优化.在交互式方法中，目标的优先权决策与Pareto解的搜索过程交替进行，变化的优先权可产生变化的非支配最优解集，决策者从搜索过程中提取有利于精炼优先权设置的信息,而优先权的设置则有利于搜索到决策者更感兴趣的非支配最优区域.后验方式是算法首先搜索到尽可能多的 Pareto 最优解,然后决策者从搜索到的解集中进行选择,选择过程是一个优先权的决策过程.

此外,如果按照适应值赋值方法分类,多目标进化算法可以分为基于 Pareto 最优概念和不基于 Pareto 最优概念两类方法.

从统计数据中发现,目前比较流行后验方法,数量几乎是另外两种的两倍之多;在后验方法中,使用 Pareto 选择策略的又是其他选择策略的两倍多.因此,目前基于 Pareto 的进化方法在多目标进化算法设计和应用中占据主导地位.

(2) 多目标进化算法的研究方向

纵观 MOEA 的研究成果,大部分还是集中在算法的设计上,对算法的性能、收敛性分析等理论性的研究则很少,偶有一些理论成果也仅仅局限于对算法的参数、状态等研究之上,且内容和深度都较浅,因此理论研究大大滞后于 MOEA 在工程中的应用.

对 MOEA 的研究主要是两个方面:其一,如何避免未成熟收敛,即保持非劣解向 Pareto 界面移动;其二,使获得的 Pareto 最优解(非劣解) 均匀分布且范围最广,即保持解的多样性.

(3) 多目标优化遗传算法

在过去的 20 年中,遗传算法作为一种好的进化算法在多目标优化问题的求解方面受到了相当程度的关注，这就诞生了遗传多目标优化.这个主题已经由Fonseca和Fleming，Hom，Tamaki，Kita和Kobayashi进行了综述，此处简单介绍一下这方面的内容.

① 遗传算法的发展过程

遗传算法是一种有效的求解优化问题的算法，最先由Michigan大学教授John Holland提出.1975年他的*Adaptation in Natural and Artificial System*一书问世标志着遗传算法的诞生，随后发展成一种通过模拟自然进化过程解决最优化问题的计算模型.De Jong在其博士论文中结合模式定理进行了大量的纯数值函数优化测试平台，解决了遗传算法的测试问题，同时，还树立了遗传算法的工作框架，得到了一些重要且有指导意义的结论，为GA的广泛应用奠定了坚实的基础.1989年，Goldberg出版了专著《搜索、优化和机器学习中的遗传算法》(*Genetic Algorithms in Search，Optimization and Machine Learning*)，全面而完整地论述了遗传算法的基本原理及其应用，可以说奠定了现代遗传算法的科学基础，为众多研究和发展遗传算法的学者所瞩目.1991年，Davis编辑出版了《遗传算法手册》(*Handbook of Genetic Algorithms*)一书，包括了遗传算法在科学计算、工程技术和社会经济中的大量应用实例，为推广和普及遗传算法的应用起到了重要的指导作用.1985年，在美国召开了第一届遗传算法国际会议，即ICGA：(International Conference on Genetic Algorithm)，此后每隔一年举行一次.GA已成为一个多学科、多领域的重要研究方向,成为国际学术界所瞩目的热点.近几年国内许多学者也开始遗传算法的研究与应用,并获得了很多成果.

图1为遗传算法的基本流程图.

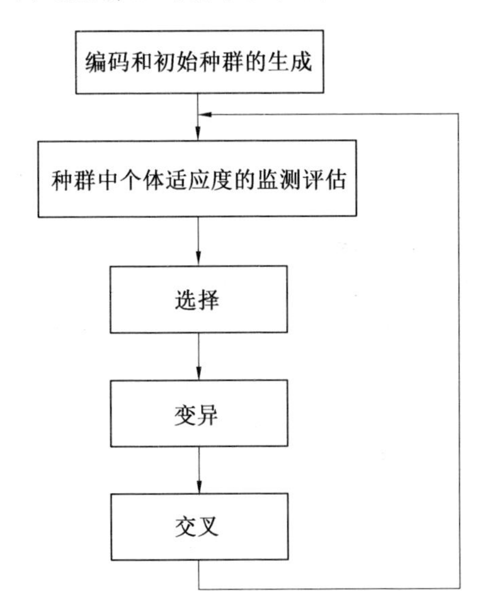

图1　遗传算法的基本流程

② 遗传算法在多目标优化中的应用

对于求解多目标优化问题的最优解,目前已有多种基于遗传算法的求解方法.下面简要介绍五种常用的方法.

（Ⅰ）权重系数变换法

对于一个多目标优化问题，若给其每个子目标函数 $f(x_i)(i=1,2,\cdots,n)$ 赋予权重 $\lambda_i(i=1,2,\cdots,n)$，其中 λ_i 为相应的 $f(x_i)$ 在多目标优化问题中的重要程度，则各个子目标函数 $f(x_i)$ 的线性加权和表示为 $\mu=\sum_{i=1}^{n}\lambda_i\cdot f(x_i)$.

若将 μ 作为多目标优化问题的评价函数，则多目标优化问题就可转化为单目标优化问题，即可以利用单目标优化的遗传算法求解多目标优化问题.

（Ⅱ）并列选择法

并列选择法的基本思想是，先将群体中的全部个体按子目标函数的数目均等地划分为一些子群体，对每个子群体分配一个子目标函数，各个子目标函数在相应的子群体中独立地进行选择运算，各自选择出一个适应度高的个体组成一个新的子群体，然后再将所有这些新生成的子群体合并成一个完整的群体，在这个群体中进行交叉和变异运算，从而生成下一代的完整群体，如此不断地进行“分割—并列选择—合并”操作，最终可求出多目标优化问题 Pareto 最优解.

（Ⅲ）排列选择法

排列选择法的基本思想是：基于 Pareto 最优个体，对群体中的各个个体进行排序，依据这个排列次序来进行进化过程中的选择运算，从而使得排在前面的 Pareto 最优个体将更多的机会遗传到下一代群体中. 如此这样经过一定代数的循环之后，最终就可求出多目标最优化问题的 Pareto 最优解.

(Ⅳ) 共享函数法

求解多目标最优化问题时，一般希望所得到的解能够尽可能地分散在整个 Pareto 最优解集合中，而不是集中在其 Pareto 最优解集合内的某一个较小的区域上. 为达到这个要求，可以利用小生境遗传算法的技术来求解多目标最优化问题，这种方法称为共享函数(Sharing Function)法，它将共享函数的概念引入到求解多目标最优化问题的遗传算法中. 算法对相同的个体或类似个体的数量加以限制，以便能够产生出种类较多的不同的最优解.

对于一个个体 X，在它的附近还存在有多少种、多大程度相似的个体，是可以度量的，这种度量值称为小生境数，小生境数的计算方法定义为

$$m_X = \sum_{Y \leqslant n} s[d(X,Y)]$$

式中，$s(d)$ 为共享函数，它是个体之间距离 d 的单调递减函数. $d(X,Y)$ 可以定义为个体 X,Y 之间的海明距离.

在计算出各个个体的小生境数之后，可以使小生境数较小的个体能够有更多的机会被选中，遗传到下一代群体中，即相似程度较小的个体能够有更多的机会被遗传到下一代群体中，这样就增加了群体的多样性，也增加了解的多样性.

(Ⅴ) 混合法

混合法的基本思想，选择算子的主体使用并列选择法，然后通过引入保留最佳个体和共享函数的思想来弥

补只使用并列选择法的不足之处.算法的主要过程为:

(i) 并列选择过程.按所求多目标优化问题的子目标函数的个数,将整个群体均等地划分成一些子群体,各个子目标函数在相应的子群体中产生其下一代子群体.

(ii) 保留Pareto最优个体过程.对于子群体中的Pareto最优个体,不让其参与个体的交叉运算和变异运算,而是将这个或这些Pareto最优个体直接保留到下一代子群体中.

(iii) 共享函数处理过程.若所得到的Pareto最优个体的数量已经超过规定的群体规模,则需要利用共享函数的处理方法来对这些Pareto最优个体进行挑选,以形成规定规模的新一代群体.

③ 遗传算法的优缺点

遗传算法的优越性可以简单地归结为以下三条:

(Ⅰ) 遗传算法适合数值求解那些带有多参数、多变量、多目标和在多区域但连通性较差的NP难优化问题.对多参数、多变量的NP难优化问题,通过解析求解或是计算求最优解的可能性很小,主要依赖于数值求解.遗传算法是一种数值求解的方法,具有普适性,对目标函数的性质几乎没有要求,甚至都不一定显式地写出目标函数,因此用遗传算法求最优化问题不足为奇.遗传算法所具有的特点是记录一个群体,它可以记录多个解而不同于局部搜索、禁忌搜索和模拟退后仅仅是一个解,这多个解的进化过程正好适合于多目标最优化问题的求解.

(Ⅱ) 遗传算法在求解组合优化问题时,不需要很强的技巧和对问题有非常深刻的理解.如排序、路线调度问题、布局问题,等等.遗传算法在给这些问题的决策变量编码后,其计算过程是比较简单的,且可以较快得到一个满意解.

(Ⅲ)遗传算法同求解问题的其他启发式算法有很好的兼容性.如可以用其他的算法求初始解;在每一个群体,可以用其他方法求解下一代新群体.

遗传算法也不可避免地存在着它的不足之处:

(i) 存在编码不规范及表述不准确等问题.

(ii) 单一的遗传算法编码不能全面地将优化问题的约束表示出来.考虑约束的一个方法就是对不可行解采用阈值,这样计算的时间必然增加.

(iii) 遗传算法通常的效率比其他传统的优化方法低.

(iv) 遗传算法容易导致出现过早收敛.

(v) 遗传算法对算法的精度、可信度、计算复杂性等方面,还没有有效的定量分析方法.

④ 遗传算法的改进

遗传算法虽然在过去的20年中得到了广泛的应用,但研究人员已经意识到,遗传算法采用简单的、固定不变的进化策略对复杂应用场合的效果并不理想,传统的遗传算法逐渐暴露出一些缺点.所以,为了提高遗传算法的性能,使其更好地应用于实际问题的解决中,研究者们开始对遗传算法进行改进,其基本途径概括起来主要有下面几个方面:

(i) 改进遗传算法的组成成分或使用技术，如选用优化控制参数、适合问题特性的编码技术等.

(ii) 采用混合遗传算法(Hybrid Genetic Algorithm).

(iii) 采用动态自适应技术，在进化过程中调整算法控制参数和编码精度.

(iv) 采用非标准的遗传操作算子.

(v) 采用并行算法.

在许多资料中还介绍了七种改进的遗传算法：

(i) 分层遗传算法(Hierarchic Genetic Algorithm)；

(ii)CHC 算法；

(iii)Messy 遗传算法；

(iv) 自适应遗传算法(Adaptive Genetic Algorithm)；

(v) 基于小生境技术的遗传算法(Niched Genetic Algorithm)；

(vi) 并行遗传算法(Parallel Genetic Algorithm)；

(vii) 混合遗传算法：遗传算法和最速下降法结合遗传算法；遗传算法和模拟退火法相结合的混合遗传算法.

图 2 为改进遗传算法的程序流程图.

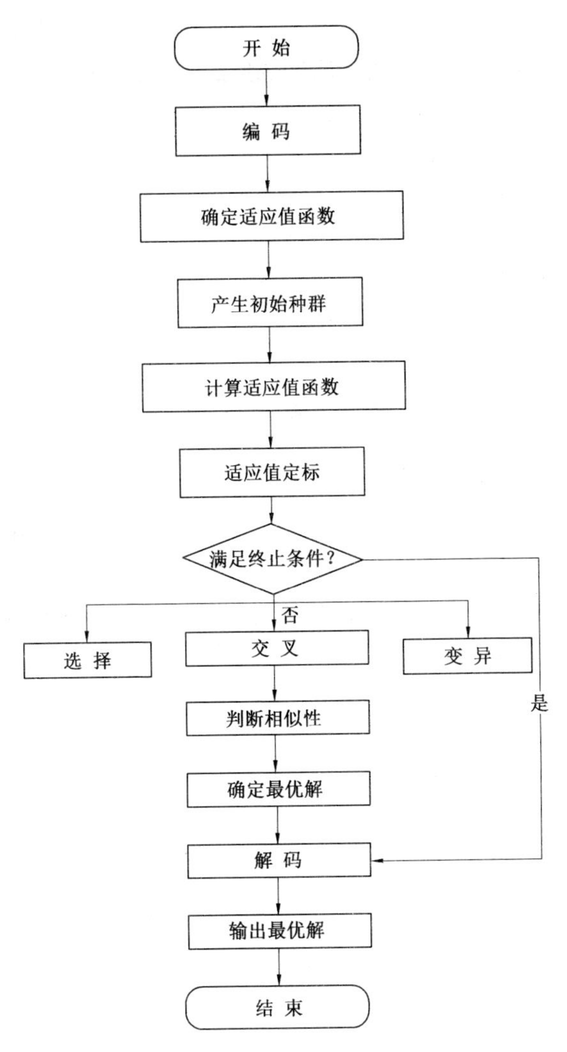

图 2　改进遗传算法的程序流程

据统计，从1990年到2008年，国内已发表这方面的文章1 600多篇，很多优秀的硕士、博士论文也从这方面选题，遗传算法的这些扩展和改进给一般问题特别是工业工程中难以求解的优化问题带来了新的希望和方向.

① 遗传算法与传统优化算法的比较

为解决各种优化计算问题，人们提出了各种各样的优化算法，如单纯型法、梯度法、动态规划法、分支定界法等.这些优化算法各有各的长处，各有各的适用范围，也各有各的限制.如寻找的是局部最优解、需要目标函数连续可微、计算量大、搜索效率低等问题.而GA采用了独特的搜索技术，克服了传统搜索方法存在的不足.

与传统方法相比，GA具有以下特点：

（Ⅰ）遗传算法以决策变量的编码作为运算对象.传统的优化算法往往直接用变量的实际值本身进行优化计算，但遗传算法不是直接以变量的值，而是以变量的某种形式的编码为运算对象.尤其对一些无数值概念或很难有数值概念，而只有代码概念的优化问题，编码处理方式更显示了其独特的优越性.

（Ⅱ）遗传算法直接以目标函数值作为搜索信息.传统的优化算法不仅需要利用目标函数值，而且往往需要目标函数的导数值等其他一些辅助信息才能确定搜索方向.而遗传算法仅使用由目标函数值变换来的适应度函数值，就可确定进一步的搜索方向和搜索范围.这个特性对很多目标函数是无法或很难求导数的函数，或导数不存在的函数的优化问题，以及组合优化问题等，应用遗传算法显得比较方便.再者,直接使用目标函数值或个体适应度,也可使得我们可以把搜索范围集中到适应高的部分搜索空间中,从而提高了搜索效率.

(Ⅲ)GA采用了多点搜索技术.GA同时对解空间的不同区域进行搜索,这是遗传算法所特有的一种隐含并行性,可避免搜索过程陷入局部极值点中,并确保以较大的概率求得全局最优解.

(Ⅳ)遗传算法使用多点搜索技术,属于一种自适应概率搜索技术,以概率作为启发式搜索信息,因而比随机搜索法具有更高的搜索效率.

(4) 其他多目标进化算法

① 多目标粒子群算法

粒子群优化算法最早是由美国学者Kenney和Eberthart于1995年提出的,但直到2002年才逐渐应用到多目标优化问题中.Hu和Eberthart提出了一种基于动态邻居策略的多目标粒子群算法,采用一维优化来处理多个目标;Fieldsend和Singh提出了一种“支配树”的数据结构,用其来存储精英解以及为粒子选取最优经验,指导粒子飞行;Mostaghim和Lechuga提出了一种基于网格的多目标粒子群算法,将目标空间分成若干网格,随机地从含精英较少的网格中选取一个解作为当前粒子的全局最优经验.使用粒子群算法来求解多目标优化问题的研究,目前仍处于起步阶段,开展的不够广泛,有许多理论问题有待于进一步的研究和探讨.

② 神经网络、模糊理论在多目标优化中的应用

神经网络的发展已有较长的历史，也获得了广泛的应用．目前神经网络在多目标优化的应用多为和其他优化方法的结合，如和模糊理论结合构造评价准则、取代单目标优化方法．并不是在多目标优化策略方面的应用．

神经网络在多目标优化中有很多应用，如可用均匀设计和正交设计的方法选取权重系数，并将权重系数和优化结果作为训练样本，建立函数关系，这样改变权重系数后，可用网络结构求出最优解．也可以根据神经网络有很强的并行计算能力的特点，应用在分层优化中，或和遗传算法结合，从非劣解集中选择最优解．另外，也可以把模糊理论和神经网络结合起来，形成即能处理模糊信息，又有并行处理、进行学习的特点．

多目标最优解同各子目标最优解密切相关，但子目标之间，各子目标最优解同多目标最优解之间的关系是模糊的．所以用模糊优化方法往往能得到满意的结果．

模糊方法应用于多目标优化主要有以下途径：

其一是构造模糊性的多目标优化算法．模糊性多目标优化算法的基本原理为在各单目标最优解的模糊集中寻求各个目标都尽可能优的满意解．

其二用于非劣解的模糊评价．利用模糊数学的方法认为各非劣解之间及非劣解和理想解之间的关系具有模糊性的前提下，用模糊数学方法来确定最优解．

(5) 对多目标进化算法的评价

随着多目标进化算法应用于各个领域，对它的研究也将直接受到实际工程问题的推动．目前，多目标优化技术仍面临着很多困难，存在很多有待研究的问题，如非劣最优解的质量评估、多目标优化进化动力学行为及其稳定性分析、选种配对机制、适应值赋值方法、种群更新终止条件及其稳定性分析，以及实际多目标优化问题的进化求解等，这些都是目前的研究热点.

作业调度问题(Job-Shop Scheduling Problem，简称 JSP) 是工厂中常见的问题，它要求在有限资源的条件下由机器完成不同的任务，每项任务有一预定的排列顺序，一台机器一次只能执行一项任务. 要有效地完成这些任务和作业，生产车间必须确定各项任务的开始时间和结束时间，JSP 问题实际上就是要解决如何安排各项任务的时间先后，从而合理的利用有限的资源，使目标函数值最小. 琼州大学物理系的林雄、黄槐仁、张福金三位教授曾写文章介绍过遗传算法和作业调度问题，然后给出一个用遗传算法求解作业调度问题的仿真结果①.

1. 引言

遗传算法(GA) 是建立在生物界自然选择和进化机制基础上的人工智能搜索算法，传统的搜索技术在复杂的搜索空间寻找问题的解时太慢，以至于在超级计算机上才能实现. 与传统的搜索技术不同，遗传算法使用了群体搜索技术，将种群代表一组问题解，通过对当前种群施加选择、交叉和变异等一系列遗传操作，从而产生新一代

① 摘自《系统仿真技术及其应用，第 9 卷》.

的种群，并逐步使种群进化到包含近似最优解的状态. 由于其思想简单、易于实现以及表现出来的健壮性，使它在许多领域得到应用，特别近年来，在问题求解、优化和搜索、机器学习、智能控制、模式识别和人工生命等领域取得了令人瞩目的成果. 作业调度问题对不同的任务和作业要求合理安排和分派以便能在有限的资源条件下完成，它的目标是确保产品约束和生产产品的代价最少. 用遗传算法研究作业调度问题是近年来广大科技工作者广泛采用的方法.

2. 遗传算法①

传统的优化方法建立在某一函数可微的基础之上，不幸的是，真实世界中的许多问题都是不可定义的，即使被定义，使用梯度搜索方法也不能找到全局最优解，克服这一问题的有效方法是使用遗传算法. 一般来说，遗传算法由问题求解、生成新解的遗传算子和适应度函数通过编码求解方式完成. 下列程序段给出了用遗传算法进行问题求解的伪代码.

```
begin
    初始化时间 t,
    初始化 t 时刻的种群,
    求 t 时刻种群的适应度函数,
    While (世代数 < 给定的总数) do
  begin
```

① 邵军力，等. 人工智能基础. 北京：电子工业出版社，2000：203-204.

```
    从t时刻的种群中选择t+1的种群,
    对t+1时刻的种群使用交叉算子,
    对t+1时刻的种群使用变异算子,
    求t+1时刻种群的适应度函数,
    将t+1时刻当成新的时刻t+1,
  end
end
```

通常,遗传算法采用二进制或十进制编码的染色体串表示特解问题的候选解,用适应度函数评价染色体的优劣,并在此基础上进行选择、交叉和变异等遗传操作,因此适应度函数选择是否恰当对算法的性能好坏影响很大,应根据实际问题的特性具体确定.此外,遗传算法适应度函数非负,同时要求把待解优化问题表示为最大化问题,即目标函数的优化方向对应适应度函数的增大方向,而一般的优化问题并不一定满足这些条件.许多问题求解的目标是求目标函数的最小值或负值,若种群的规模不大,在遗传算法运行的开始阶段,有少数适应度极高的染色体,则按通常的选择法,其选择概率很高,这些染色体就会大量繁殖(复制),在种群中占有很大的比例,减少了种群的多样性,导致算法过早收敛,可能丢失一些最优点,而陷入局部最优点.为了解决这一问题,应使用交叉和变异等遗传算子.

3. 作业调度问题①

(1)JSP 问题的图示及选择操作

表1　3×3的JSP问题

作业	工序1 (时间)	工序2 (时间)	工序3 (时间)
1	1(3)	2(3)	3(3)
2	1(2)	3(3)	2(4)
3	2(3)	1(2)	3(1)

作业调度问题由许多机器 M 和作业 J 组成，每项作业又由具有确定时间的小作业(任务)组成，每项任务必须在指定的机器上加工，每项作业每次只能使用一台机器.在每项任务有一预定的排列顺序，设在第 r 台机器 Mr 上进行第 j 项任务 Jj 的作业是 Ojr，所需要的时间是 Pjr，一次调度就是满足这些约束的每个作业 $\{O_{jr}\}(1\leqslant j\leqslant n,1\leqslant r\leqslant m)$ 的完成时间，从第一项任务开始到最后一项任务结束所需要的时间称为遍历时间 L(Makespan)，一个调度也就是按先后顺序条件将所有任务安排到机器上进行加工的一种方案，因此，我们的目标是给每个变量寻找满足约束的确定的调度，使目标函数 L 最小化.表1给出了3台机器和3个作业的JSP问题的加工顺序(工序)及时间.

JSP问题是一NP难题，随着作业数的增多，问题变得

① BALAS E. Machine Sequencing via Disjunctive Hraphs: An Implicit Enumeration Algorithm. Operations Research, 1969(17): 941-957.

越来越复杂.

JSP 问题可用带有双向箭头有向图 $G=(V,C\cup D)$ 表示,图中的结点表示作业,0 和 * 号结点是两个假设的结点,分别表示作业的开始和结束,V 表示所有结点(含假设结点)的集合,C 表示作业加工顺序的边集,带双向箭头的虚线边集 D 表示在同一台机器上加工的两个作业,结点旁的 Pjr 表示完成该作业所需的时间. 图 1 是上述问题的图示.

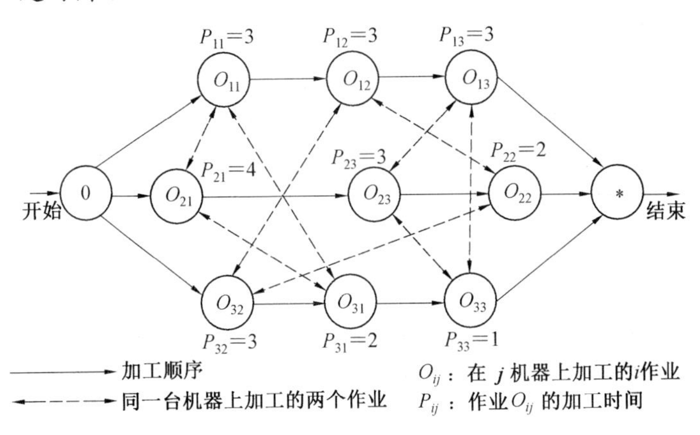

图 1 3 × 3 JSP 问题的有向图示

JSP 问题也可以看成在同一台机器上加工的全部作业的排序. 这样,在有向图模型中,我们所做的只需把带双向箭头的虚线改变成带单箭头的实线,一次选择就是从带双向箭头的虚线中选择一组带单箭头的实线,如果选择完所有的带双向箭头的虚线,则一次选择完成,这时的图是有向无环图.

(2) 交叉操作①

选择操作虽能够从旧种群中选择出优胜者,但不能创造出新的染色体,要创造出新的染色体必须用遗传操作交叉完成.以单点交叉为例,任意挑选经过选择操作后种群中两个个体作为交叉对象,即两个父个体经过染色体交换重组产生两个子个体,如图2所示.随机产生一个交叉点位置,父个体1和父个体2在交叉点位置之右的部分基因码互换,生成子个体1和子个体2.类似地完成其他个体的交叉操作.

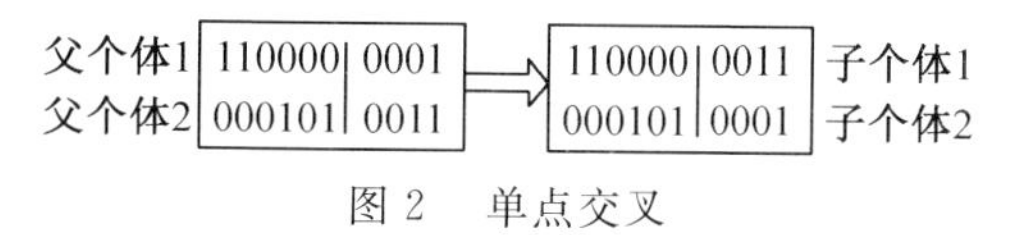

图2　单点交叉

(3) 变异操作

对于两个作业排列 $P0$ 和 $P1$,随机选择两列互换产生新的排列 $k0$ 和 $k1$,由于这样的操作产生的新的排列除交工期约束外其他约束都能满足,所以,在不能满足交工期约束时,可以再随机选择其他两列来互换产生新的排列,直到满足交工期约束.

4. 仿真结果②

用遗传算法仿真某一家模具工厂生产模具的情况,此模具由六个零件组成,而每个零件包括六道工序.以表

---

① 林雄,等.遗传算法及其进化硬件设计研究.微计算机信息,2004,2:28-30.

② Http://www.metavistas.com.tw/gademo.jsp.

2,表3说明机器与零件之间的关系.表2为零件在各台机器上的加工顺序($M[j,m]$),由于每个零件加工的顺序不同,零件或先钻或刨等根据其特性有所限制,故表2是不可变动的.表3为每个零件在每台机器上加工的时间($T[m,j]$),各零件在机器上加工的时间也是不能改变的.

表2　零件在各台机器上的加工顺序表($M[j,m]$)

	工序1	工序2	工序3	工序4	工序5	工序6
零件1	3	1	2	4	6	5
零件2	2	3	5	6	1	4
零件3	3	4	6	1	2	5
零件4	2	1	3	4	5	6
零件5	3	2	5	6	1	4
零件6	2	4	6	1	5	3

表3　零件在每台机器上加工的时间($T[m,j]$)

	零件1	零件2	零件3	零件4	零件5	零件6
机器1	3	10	9	5	3	10
机器2	6	8	1	5	3	3
机器3	1	5	5	5	9	1
机器4	7	4	4	3	1	3
机器5	6	10	7	8	5	4
机器6	3	10	8	9	4	9

表2中的行代表机器的号码,即该工序需在哪一台机器上加工.例如第1行表示:零件1的加工顺序为机器3、机器1、机器2、机器4、机器6、机器5,其余类推.

表3的行代表机器1～6,列代表零件1～6.所以第

1 行的意义代表:零件 1 在机器 1 加工 3 个单位时间,零件 2 在机器 1 加工 10 个单位时间,零件 3 在机器 1 加工 9 个单位时间,零件 4 在机器 1 加工 5 个单位时间,零件 5 在机器 1 加工 3 个单位时间,零件 6 在机器 1 加工 10 个单位时间,其余依此类推.

对上述实例的作业调度问题,用 Java Applet 编程模拟,实验中所选择的遗传算法参数列表如表 4 所示,经过世代交替后,求得的遍历时间慢慢递减,与理论相符,如图 3.

图 4 为甘特图,其中纵轴为机器编号;横轴为时间.图中的条纹方块表示零件在机器上的加工顺序.

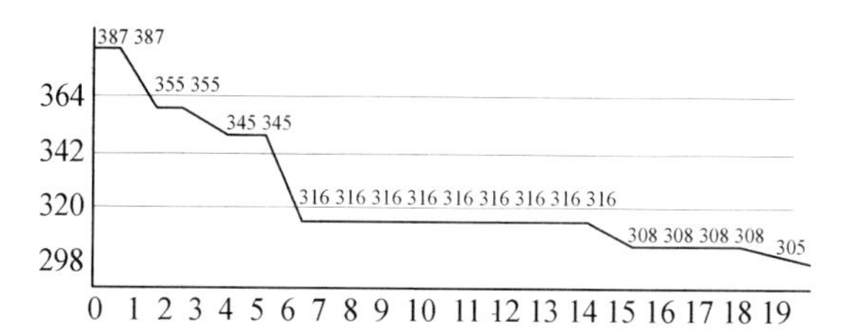

图 3　遍历时间随世代数的变化情况

表 4　遗传基因算法输入参数列表

参数名	参数的大小
种群大小	100
交叉率	0.8
变异率	0.1
世代数	100

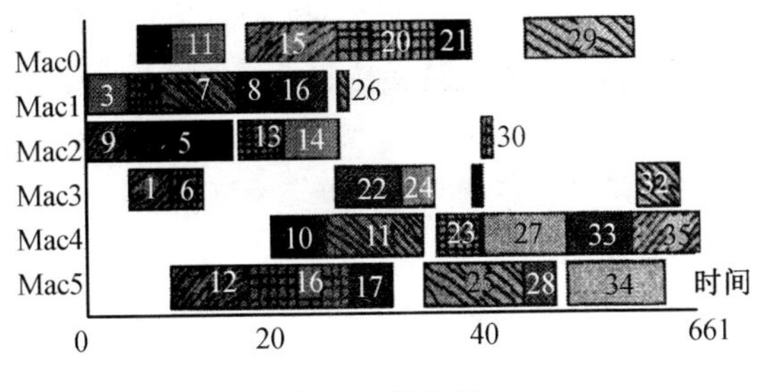

图 4　甘特图

5. **结束语**

本文分析和简要介绍了遗传算法和作业调度问题，在设计算法时以随机方式产生许多点，同时搜寻最优解，因为在每一次迭代过程皆是取相对最佳的点，因此能找到最接近之最优解. 仿真结果表明，用本文提出的遗传算法求解作业调度问题比用其他方法更优.

作业调度问题属于典型的困难组合优化问题(NP-hard)，尚没有一个有效的算法能在多项式时间内求出其最优解，一些启发式算法就成为有吸引力的备选方法①②.

在生产作业管理中引入智能主体(agent)，可以很方便地模拟和描述生产作业系统各组成要素的行为和运行机制. 基于多智能主体

① ROADAMMER F A, WHITE K P. A Recent Survey of Production Scheduling. IEEE Trans. on SMC, 1998, 18: 841-851.

② 熊锐，吴澄. 车间生产调度问题的技术现状与发展趋势. 清华大学学报(自然科学版), 1998, 38(10): 55-60.

(Multiagent)技术的合作求解方法是较新的智能调度方法，通过智能主体之间的合作以及多智能主体系统协调来完成作业任务的调度，并达到预先规定的作业目标及状态. 在这种研究方法中，在智能主体内部也可采用基于规则及智能推理相结合的混合方法来构造基于多智能主体系统的智能调度系统①②.

华中科技大学现代化管理研究所的傅小华、黎志成两位研究员于2002年针对作业调度问题探讨了基于多智能主体技术的合作求解机制，提出合作求解算法. ③

1. 基于多智能主体的合作求解机制

(1) 合作求解与冲突

在生产作业管理中引入智能主体，由于生产作业调度是通过智能主体之间的合作以及多智能主体系统协调实现的，可较好地解决不同目标之间的冲突并降低问题的复杂性，达到预先规定的目标与状态.

但是，由于多智能主体系统按功能或目标分解成若干具有独立功能的单一智能主体，各智能主体之间既相互独立又相互联系，各智能主体对各自问题的描述不同、求解策略、考虑角度、评价准则不一致，必然导致问题求解中冲突的产生. 作业调度方案的合作求解过程，也是冲

① Rabclo R J. Camarnha-Matos L M, Afsarmancsh H. Multi-agent-based agile scheduling. Robotics and Autonomous Systems, 1999, 27: 15-28.

② 史忠植. 智能主体及其应用. 北京：科学出版社，2000.

③ 摘自《华中科技大学学报(自然科学版)》，2002年，第30卷第5期.

突产生、协商与消解的过程.

(2) 冲突产生的原因与种类

冲突产生的根本原因是实际作业调度问题的复杂、多变和多目标性.在基于多智能主体的智能调度系统中冲突的产生是各智能主体的目标和效用不一致的必然结果.

根据冲突产生的原因可将冲突分为3类:a.目标冲突.由于作业调度问题的多目标性,导致智能主体的目标之间的冲突.b.工艺冲突.各智能主体在完成同一工件的加工任务时出现的冲突即违反工艺约束,最典型的就是"死锁".c.资源冲突.在同一时间,多个工件同时竞争一个生产资源的冲突.

第一类冲突属于目标冲突,应使用目标冲突消解规则解决,即如果一个智能主体有两个冲突的目标,那么选择效用大的目标,删除效用小的目标.后两种冲突属于规划冲突,应使用规划选择规则加以解决,即如果智能主体有可以实现目标 r 的两个规划,那么选择效用大或代价小的规划.

(3) 基于知识的协商消解冲突方式

在基于多智能主体的系统中解决冲突的方式有多种,如回溯、约束松弛、中止合作及协商消解等方式①.回溯、约束松弛与中止合作等方式都存在危及多智能主体系统结构及最终目标实现的可能,而协商消解冲突最有

① 史忠植.智能主体及其应用.北京:科学出版社,2000.

利于多智能主体系统最终目标的实现，因此，协商消解冲突已成为多智能主体系统合作求解的最主要的策略.

作业调度问题是一个典型的面向共同目标的合作求解问题，各智能主体之间是强依赖关系，缺少任何一个智能主体的参与都会危及问题的求解，因此解决冲突(不论目标冲突还是规划冲突)时必须采取以让步为基本特征的协商消解冲突方式.

① 协商

协商是智能主体之间寻求在尽可能既有利于自身目标，又帮助其他智能主体改善目标，使整体目标优化的折中与让步行为.

协商可以发生在双方之间，也可发生在多方之间.为简化问题，可认为多方协商是多次双方协商的综合结果.从而将关注的焦点放在双方协商上.

双方协商的结果可能有以下3种：a.协商各方的目标都有所改善；b.协商各方的目标都有所损失；c.协商一方的目标改善，而另一方目标有所损失.

协商结果能否接受取决于整体目标是否为可选择诸方案中最佳.

② 基于知识的协商消解冲突

生产作业调度的研究不论在理论上，还是在实践上都取得了不少成果，完全可以将它们应用于多智能主体之间的冲突消解.

所谓基于知识的协商消解冲突就是将生产作业调度领域内的知识和过去行之有效的经验用于冲突的探测、

识别、协调方案的构造以及协商结果的评价.

a. 目标冲突,应使用目标冲突消解规则解决,根据理论知识和实践经验判定各目标的重要程度,设置相应的权值,计算智能主体针对不同目标 r_1 和 r_2 的目标效用值 u_1 和 u_2,决定解决方案,当智能主体上的所有目标冲突都消解了,智能主体的意图和规划也就相应确定了;

b. 工艺冲突和资源冲突,属于规划冲突,应按照规划选择规则加以解决,根据理论知识和实践经验加以识别并构造协商解决方案,根据整体目标效用值决定解决方案的抉择.

对于作业调度问题,各智能主体的规划是全局规划的一个组成部分,全局规划中存在的冲突可分解智能主体之间的两两冲突,解决了智能主体之间的冲突也就解决了全局规划的求解.

对于智能主体1和2,在出现规划冲突时,根据各自目标效用及相关知识构造最有利于自己并易于为对方接受的解决方案(规划)P_{L1} 和 P_{L2},计算出相应的目标效用变化值 u_1 和 u_2,再按规划选择规则决定一方案为两个智能主体的合作(子)规划

$$P(1,P_{L1},r,u_1) \wedge P(2,P_{L2},r,u_2) \wedge (u_1 \geqslant u_2) \Rightarrow P(1,P_{L1},r,u_1)$$

$$P(1,P_{L1},r,u_1) \wedge P(2,P_{L2},r,u_2) \wedge (u_1 \leqslant u_2) \Rightarrow P(2,P_{L1},r,u_1)$$

由于基于知识的协商消解冲突的解决方案都是在满足整体目标及其效用值最大化的基础上确定的,因此,对

生产作业系统整体而言是合理的；

同时，各智能主体的目标及效用值只有在保证整体目标的基础上才会有所牺牲，它是在保证整体目标及效用情况下所得到的最为有利的解决方案.

所以，建立在基于知识的协商基础上的合作求解机制是有效且可行的.

2. 智能主体的结构模型与合作求解算法

根据上面的讨论，可以构造智能主体的基本结构和以让步协商为基础的具体的合作求解算法.

(1) 智能主体的结构模型

基于多智能主体的智能调度系统由多个类似的智能主体组成，每个智能主体对应生产作业系统中的一个生产加工资源.

单个智能主体拥有三个执行模块和四个知识模块，智能主体的结构如图1所示.

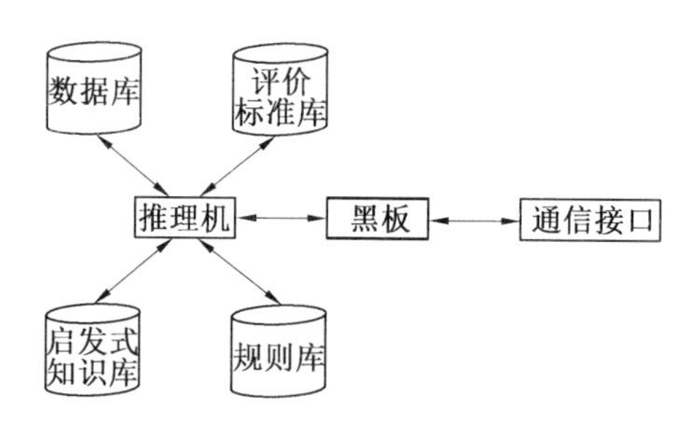

图1　智能主体结构

数据库. 存放作业调度问题的基本数据，是可能与可达世界空间的体现，与信念(Belief) 可达关系相对应.

评价标准库. 存放作业调度问题解空间的评价指标与标准,包括各类相关因素的相对重要程度判断标准与权重的设置方法,是可达世界状态空间及其转变方式——喜好的体现,与愿望(Desire)可达关系相对应.

规则库. 存放求解作业调度问题的必然性知识,包括工序约束、资源约束、解析算法等,即获得解空间过程中有关状态转变的必然性知识.

启发式知识库. 存放求解作业调度问题的或然性知识,包括各种优先调度规则、启发式算法等,即获得解空间过程中有关状态转变的或然性知识.

规则库与启发式知识库共同与意图(Intention)可达关系相对应.

推理机. 根据四个知识模块的知识进行推理:识别冲突、得到规划及给出智能主体执行该规划导致的价值量的变化;在此约定采用正向推理策略.

黑板. 用于存放协商过程中的中间信息.

通信接口. 实现智能主体之间的交互.

(2) 合作求解算法

基于多智能主体技术的合作求解算法的基本思想:各智能主体根据自身目标对相关作业任务进行安排,从而形成初始解(规划),该初始解满足各智能主体目标及效用最大化,但可能不可行(即存在各类冲突),需在协商中逐步让步(降低自身目标效用),使冲突不断消解,最终达到可行,该规划即为在可行条件下对系统整体及各智能主体最为有利的解. 具体步骤如下:

步骤1 令$k=0$,各生产智能主体根据目标或效用函数生成本生产智能主体子调度S_i,形成初始调度方案$S_0(S_1,S_2,\cdots,S_n)$.

步骤2 对于调度方案$S_k(S_1,S_2,\cdots,S_n)$,各生产智能主体计算本智能主体的目标或效用值$G_k(S_i)$.

步骤3 判断调度方案$S_k(S_1,S_2,\cdots,S_n)$是否存在冲突:不存在,转步骤5;存在,计算冲突智能主体对A和B调整的目标或效用值减少量$\Delta G_k(S_A)$和$\Delta G_k(S_B)$.

步骤4 若$\Delta G_k(S_A)>\Delta G_k(S_B)$,调整冲突智能主体A,否则调整B,令$k=k+1$,形成调度方案$S_{k+1}(S_1,S_2,\cdots,S_n)$,转步骤2.

步骤5 输出调度规划$S_k(S_1,S_2,\cdots,S_n)$,结束.

本数学工作室所出版的许多图书都得到了读者的认可,也有一些自视甚高者不以为然.有句话经常用来回怼,那就是:“You can you up,你行你上啊!”

刘培杰

2023.1.24

于哈工大

刘培杰数学工作室
已出版(即将出版)图书目录——原版影印

书　　名	出版时间	定　价	编号
数学物理大百科全书.第1卷(英文)	2016—01	418.00	508
数学物理大百科全书.第2卷(英文)	2016—01	408.00	509
数学物理大百科全书.第3卷(英文)	2016—01	396.00	510
数学物理大百科全书.第4卷(英文)	2016—01	408.00	511
数学物理大百科全书.第5卷(英文)	2016—01	368.00	512
zeta 函数,q-zeta 函数,相伴级数与积分(英文)	2015—08	88.00	513
微分形式:理论与练习(英文)	2015—08	58.00	514
离散与微分包含的逼近和优化(英文)	2015—08	58.00	515
艾伦·图灵:他的工作与影响(英文)	2016—01	98.00	560
测度理论概率导论,第2版(英文)	2016—01	88.00	561
带有潜在故障恢复系统的半马尔柯夫模型控制(英文)	2016—01	98.00	562
数学分析原理(英文)	2016—01	88.00	563
随机偏微分方程的有效动力学(英文)	2016—01	88.00	564
图的谱半径(英文)	2016—01	58.00	565
量子机器学习中数据挖掘的量子计算方法(英文)	2016—01	98.00	566
量子物理的非常规方法(英文)	2016—01	118.00	567
运输过程的统一非局部理论:广义波尔兹曼物理动力学,第2版(英文)	2016—01	198.00	568
量子力学与经典力学之间的联系在原子、分子及电动力学系统建模中的应用(英文)	2016—01	58.00	569
算术域(英文)	2018—01	158.00	821
高等数学竞赛:1962—1991年的米洛克斯·史怀哲竞赛(英文)	2018—01	128.00	822
用数学奥林匹克精神解决数论问题(英文)	2018—01	108.00	823
代数几何(德文)	2018—04	68.00	824
丢番图逼近论(英文)	2018—01	78.00	825
代数几何学基础教程(英文)	2018—01	98.00	826
解析数论入门课程(英文)	2018—01	78.00	827
数论中的丢番图问题(英文)	2018—01	78.00	829
数论(梦幻之旅):第五届中日数论研讨会演讲集(英文)	2018—01	68.00	830
数论新应用(英文)	2018—01	68.00	831
数论(英文)	2018—01	78.00	832

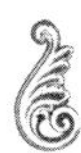

刘培杰数学工作室
已出版(即将出版)图书目录——原版影印

书　　名	出版时间	定　价	编号
湍流十讲(英文)	2018-04	108.00	886
无穷维李代数:第3版(英文)	2018-04	98.00	887
等值、不变量和对称性(英文)	2018-04	78.00	888
解析数论(英文)	2018-09	78.00	889
《数学原理》的演化:伯特兰·罗素撰写第二版时的手稿与笔记(英文)	2018-04	108.00	890
哈密尔顿数学论文集(第4卷):几何学、分析学、天文学、概率和有限差分等(英文)	2019-05	108.00	891
偏微分方程全局吸引子的特性(英文)	2018-09	108.00	979
整函数与下调和函数(英文)	2018-09	118.00	980
幂等分析(英文)	2018-09	118.00	981
李群,离散子群与不变量理论(英文)	2018-09	108.00	982
动力系统与统计力学(英文)	2018-09	118.00	983
表示论与动力系统(英文)	2018-09	118.00	984
分析学练习.第1部分(英文)	2021-01	88.00	1247
分析学练习.第2部分,非线性分析(英文)	2021-01	88.00	1248
初级统计学:循序渐进的方法:第10版(英文)	2019-05	68.00	1067
工程师与科学家微分方程用书:第4版(英文)	2019-07	58.00	1068
大学代数与三角学(英文)	2019-06	78.00	1069
培养数学能力的途径(英文)	2019-07	38.00	1070
工程师与科学家统计学:第4版(英文)	2019-06	58.00	1071
贸易与经济中的应用统计学:第6版(英文)	2019-06	58.00	1072
傅立叶级数和边值问题:第8版(英文)	2019-05	48.00	1073
通往天文学的途径:第5版(英文)	2019-05	58.00	1074
拉马努金笔记.第1卷(英文)	2019-06	165.00	1078
拉马努金笔记.第2卷(英文)	2019-06	165.00	1079
拉马努金笔记.第3卷(英文)	2019-06	165.00	1080
拉马努金笔记.第4卷(英文)	2019-06	165.00	1081
拉马努金笔记.第5卷(英文)	2019-06	165.00	1082
拉马努金遗失笔记.第1卷(英文)	2019-06	109.00	1083
拉马努金遗失笔记.第2卷(英文)	2019-06	109.00	1084
拉马努金遗失笔记.第3卷(英文)	2019-06	109.00	1085
拉马努金遗失笔记.第4卷(英文)	2019-06	109.00	1086
数论:1976年纽约洛克菲勒大学数论会议记录(英文)	2020-06	68.00	1145
数论:卡本代尔1979:1979年在南伊利诺伊卡本代尔大学举行的数论会议记录(英文)	2020-06	78.00	1146
数论:诺德韦克豪特1983:1983年在诺德韦克豪特举行的Journees Arithmetiques数论大会会议记录(英文)	2020-06	68.00	1147
数论:1985-1988年在纽约城市大学研究生院和大学中心举办的研讨会(英文)	2020-06	68.00	1148

刘培杰数学工作室
已出版(即将出版)图书目录——原版影印

书　名	出版时间	定　价	编号
数论:1987 年在乌尔姆举行的 Journees Arithmetiques 数论大会会议记录(英文)	2020—06	68.00	1149
数论:马德拉斯 1987:1987 年在马德拉斯安娜大学举行的国际拉马努金百年纪念大会会议记录(英文)	2020—06	68.00	1150
解析数论:1988 年在东京举行的日法研讨会会议记录(英文)	2020—06	68.00	1151
解析数论:2002 年在意大利切特拉罗举行的 C. I. M. E. 暑期班演讲集(英文)	2020—06	68.00	1152
量子世界中的蝴蝶:最迷人的量子分形故事(英文)	2020—06	118.00	1157
走进量子力学(英文)	2020—06	118.00	1158
计算物理学概论(英文)	2020—06	48.00	1159
物质,空间和时间的理论:量子理论(英文)	2020—10	48.00	1160
物质,空间和时间的理论:经典理论(英文)	2020—10	48.00	1161
量子场理论:解释世界的神秘背景(英文)	2020—07	38.00	1162
计算物理学概论(英文)	2020—06	48.00	1163
行星状星云(英文)	2020—10	38.00	1164
基本宇宙学:从亚里士多德的宇宙到大爆炸(英文)	2020—08	58.00	1165
数学磁流体力学(英文)	2020—07	58.00	1166
计算科学:第 1 卷,计算的科学(日文)	2020—07	88.00	1167
计算科学:第 2 卷,计算与宇宙(日文)	2020—07	88.00	1168
计算科学:第 3 卷,计算与物质(日文)	2020—07	88.00	1169
计算科学:第 4 卷,计算与生命(日文)	2020—07	88.00	1170
计算科学:第 5 卷,计算与地球环境(日文)	2020—07	88.00	1171
计算科学:第 6 卷,计算与社会(日文)	2020—07	88.00	1172
计算科学. 别卷,超级计算机(日文)	2020—07	88.00	1173
多复变函数论(日文)	2022—06	78.00	1518
复变函数入门(日文)	2022—06	78.00	1523
代数与数论:综合方法(英文)	2020—10	78.00	1185
复分析:现代函数理论第一课(英文)	2020—07	58.00	1186
斐波那契数列和卡特兰数:导论(英文)	2020—10	68.00	1187
组合推理:计数艺术介绍(英文)	2020—07	88.00	1188
二次互反律的傅里叶分析证明(英文)	2020—07	48.00	1189
旋瓦兹分布的希尔伯特变换与应用(英文)	2020—07	58.00	1190
泛函分析:巴拿赫空间理论入门(英文)	2020—07	48.00	1191
卡塔兰数入门(英文)	2019—05	68.00	1060
测度与积分(英文)	2019—04	68.00	1059
组合学手册. 第一卷(英文)	2020—06	128.00	1153
＊一代数、局部紧群和巴拿赫＊一代数丛的表示. 第一卷,群和代数的基本表示理论(英文)	2020—05	148.00	1154
电磁理论(英文)	2020—08	48.00	1193
连续介质力学中的非线性问题(英文)	2020—09	78.00	1195
多变量数学入门(英文)	2021—05	68.00	1317
偏微分方程入门(英文)	2021—05	88.00	1318
若尔当典范性:理论与实践(英文)	2021—07	68.00	1366
伽罗瓦理论. 第 4 版(英文)	2021—08	88.00	1408

刘培杰数学工作室
已出版(即将出版)图书目录——原版影印

书　名	出版时间	定　价	编号
典型群,错排与素数(英文)	2020－11	58.00	1204
李代数的表示:通过 gln 进行介绍(英文)	2020－10	38.00	1205
实分析演讲集(英文)	2020－10	38.00	1206
现代分析及其应用的课程(英文)	2020－10	58.00	1207
运动中的抛射物数学(英文)	2020－10	38.00	1208
2一纽结与它们的群(英文)	2020－10	38.00	1209
概率,策略和选择:博弈与选举中的数学(英文)	2020－11	58.00	1210
分析学引论(英文)	2020－11	58.00	1211
量子群:通往流代数的路径(英文)	2020－11	38.00	1212
集合论入门(英文)	2020－10	48.00	1213
酉反射群(英文)	2020－11	58.00	1214
探索数学:吸引人的证明方式(英文)	2020－11	58.00	1215
微分拓扑短期课程(英文)	2020－10	48.00	1216
抽象凸分析(英文)	2020－11	68.00	1222
费马大定理笔记(英文)	2021－03	48.00	1223
高斯与雅可比和(英文)	2021－03	78.00	1224
π与算术几何平均:关于解析数论和计算复杂性的研究(英文)	2021－01	58.00	1225
复分析入门(英文)	2021－03	48.00	1226
爱德华·卢卡斯与素性测定(英文)	2021－03	78.00	1227
通往凸分析及其应用的简单路径(英文)	2021－01	68.00	1229
微分几何的各个方面.第一卷(英文)	2021－01	58.00	1230
微分几何的各个方面.第二卷(英文)	2020－12	58.00	1231
微分几何的各个方面.第三卷(英文)	2020－12	58.00	1232
沃克流形几何学(英文)	2020－11	58.00	1233
彷射和韦尔几何应用(英文)	2020－12	58.00	1234
双曲几何学的旋转向量空间方法(英文)	2021－02	58.00	1235
积分:分析学的关键(英文)	2020－12	48.00	1236
为有天分的新生准备的分析学基础教材(英文)	2020－11	48.00	1237
数学不等式.第一卷.对称多项式不等式(英文)	2021－03	108.00	1273
数学不等式.第二卷.对称有理不等式与对称无理不等式(英文)	2021－03	108.00	1274
数学不等式.第三卷.循环不等式与非循环不等式(英文)	2021－03	108.00	1275
数学不等式.第四卷.Jensen 不等式的扩展与加细(英文)	2021－03	108.00	1276
数学不等式.第五卷.创建不等式与解不等式的其他方法(英文)	2021－04	108.00	1277

刘培杰数学工作室
已出版(即将出版)图书目录——原版影印

书　　名	出版时间	定　价	编号
冯·诺依曼代数中的谱位移函数:半有限冯·诺依曼代数中的谱位移函数与谱流(英文)	2021－06	98.00	1308
链接结构:关于嵌入完全图的直线中链接单形的组合结构(英文)	2021－05	58.00	1309
代数几何方法.第1卷(英文)	2021－06	68.00	1310
代数几何方法.第2卷(英文)	2021－06	68.00	1311
代数几何方法.第3卷(英文)	2021－06	58.00	1312

书　　名	出版时间	定　价	编号
代数、生物信息和机器人技术的算法问题.第四卷,独立恒等式系统(俄文)	2020－08	118.00	1199
代数、生物信息和机器人技术的算法问题.第五卷,相对覆盖性和独立可拆分恒等式系统(俄文)	2020－08	118.00	1200
代数、生物信息和机器人技术的算法问题.第六卷,恒等式和准恒等式的相等 问题、可推导性和可实现性(俄文)	2020－08	128.00	1201
分数阶微积分的应用:非局部动态过程,分数阶导热系数(俄文)	2021－01	68.00	1241
泛函分析问题与练习:第2版(俄文)	2021－01	98.00	1242
集合论、数学逻辑和算法论问题:第5版(俄文)	2021－01	98.00	1243
微分几何和拓扑短期课程(俄文)	2021－01	98.00	1244
素数规律(俄文)	2021－01	88.00	1245
无穷边值问题解的递减:无界域中的拟线性椭圆和抛物方程(俄文)	2021－01	48.00	1246
微分几何讲义(俄文)	2020－12	98.00	1253
二次型和矩阵(俄文)	2021－01	98.00	1255
积分和级数.第2卷,特殊函数(俄文)	2021－01	168.00	1258
积分和级数.第3卷,特殊函数补充:第2版(俄文)	2021－01	178.00	1264
几何图上的微分方程(俄文)	2021－01	138.00	1259
数论教程:第2版(俄文)	2021－01	98.00	1260
非阿基米德分析及其应用(俄文)	2021－03	98.00	1261
古典群和量子群的压缩(俄文)	2021－03	98.00	1263
数学分析习题集.第3卷,多元函数:第3版(俄文)	2021－03	98.00	1266
数学习题:乌拉尔国立大学数学力学系大学生奥林匹克(俄文)	2021－03	98.00	1267
柯西定理和微分方程的特解(俄文)	2021－03	98.00	1268
组合极值问题及其应用:第3版(俄文)	2021－03	98.00	1269
数学词典(俄文)	2021－01	98.00	1271
确定性混沌分析模型(俄文)	2021－06	168.00	1307
精选初等数学习题和定理.立体几何.第3版(俄文)	2021－03	68.00	1316
微分几何习题:第3版(俄文)	2021－05	98.00	1336
精选初等数学习题和定理.平面几何.第4版(俄文)	2021－05	68.00	1335
曲面理论在欧氏空间 En 中的直接表示(俄文)	2022－01	68.00	1444
维纳—霍普夫离散算子和托普利兹算子:某些可数赋范空间中的诺特性和可逆性(俄文)	2022－03	108.00	1496
Maple中的数论:数论中的计算机计算(俄文)	2022－03	88.00	1497
贝尔曼和克努特问题及其概括:加法运算的复杂性(俄文)	2022－03	138.00	1498

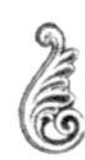

刘培杰数学工作室
已出版(即将出版)图书目录——原版影印

书 名	出版时间	定 价	编号
复分析:共形映射(俄文)	2022-07	48.00	1542
微积分代数样条和多项式及其在数值方法中的应用(俄文)	2022-08	128.00	1543
蒙特卡罗方法中的随机过程和场模型:算法和应用(俄文)	2022-08	88.00	1544
线性椭圆型方程组:论二阶椭圆型方程的迪利克雷问题(俄文)	2022-08	98.00	1561
动态系统解的增长特性:估值、稳定性、应用(俄文)	2022-08	118.00	1565
群的自由积分解:建立和应用(俄文)	2022-08	78.00	1570
混合方程和偏差自变数方程问题:解的存在和唯一性(俄文)	2023-01	78.00	1582
拟度量空间分析:存在和逼近定理(俄文)	2023-01	108.00	1583
二维和三维流形上函数的拓扑性质:函数的拓扑分类(俄文)	2023-03	68.00	1584
齐次马尔科夫过程建模的矩阵方法:此类方法能够用于不同目上的的复杂系统研究、设计和完善(俄文)	2023-03	68.00	1594

书 名	出版时间	定 价	编号
狭义相对论与广义相对论:时空与引力导论(英文)	2021-07	88.00	1319
束流物理学和粒子加速器的实践介绍:第2版(英文)	2021-07	88.00	1320
凝聚态物理中的拓扑和微分几何简介(英文)	2021-05	88.00	1321
混沌映射:动力学、分形学和快速涨落(英文)	2021-05	128.00	1322
广义相对论:黑洞、引力波和宇宙学介绍(英文)	2021-06	68.00	1323
现代分析电磁均质化(英文)	2021-06	68.00	1324
为科学家提供的基本流体动力学(英文)	2021-06	88.00	1325
视觉天文学:理解夜空的指南(英文)	2021-06	68.00	1326
物理学中的计算方法(英文)	2021-06	68.00	1327
单星的结构与演化:导论(英文)	2021-06	108.00	1328
超越居里:1903 年至 1963 年物理界四位女性及其著名发现(英文)	2021-06	68.00	1329
范德瓦尔斯流体热力学的进展(英文)	2021-06	68.00	1330
先进的托卡马克稳定性理论(英文)	2021-06	88.00	1331
经典场论导论:基本相互作用的过程(英文)	2021-07	88.00	1332
光致电离量子动力学方法原理(英文)	2021-07	108.00	1333
经典域论和应力:能量张量(英文)	2021-05	88.00	1334
非线性太赫兹光谱的概念与应用(英文)	2021-06	68.00	1337
电磁学中的无穷空间并矢格林函数(英文)	2021-06	88.00	1338
物理科学基础数学.第1卷,齐次边值问题、傅里叶方法和特殊函数(英文)	2021-07	108.00	1339
离散量子力学(英文)	2021-07	68.00	1340
核磁共振的物理学和数学(英文)	2021-07	108.00	1341
分子水平的静电学(英文)	2021-08	68.00	1342
非线性波:理论、计算机模拟、实验(英文)	2021-06	108.00	1343
石墨烯光学:经典问题的电解解决方案(英文)	2021-06	68.00	1344
超材料多元宇宙(英文)	2021-07	68.00	1345
银河系外的天体物理学(英文)	2021-07	68.00	1346
原子物理学(英文)	2021-07	68.00	1347
将光打结:将拓扑学应用于光学(英文)	2021-07	68.00	1348
电磁学:问题与解法(英文)	2021-07	88.00	1364
海浪的原理:介绍量子力学的技巧与应用(英文)	2021-07	108.00	1365
多孔介质中的流体:输运与相变(英文)	2021-07	68.00	1372
洛伦兹群的物理学(英文)	2021-08	68.00	1373
物理导论的数学方法和解决方法手册(英文)	2021-08	68.00	1374

刘培杰数学工作室
已出版(即将出版)图书目录——原版影印

书　　名	出版时间	定　价	编号
非线性波数学物理学入门(英文)	2021—08	88.00	1376
波:基本原理和动力学(英文)	2021—07	68.00	1377
光电子量子计量学.第1卷,基础(英文)	2021—07	88.00	1383
光电子量子计量学.第2卷,应用与进展(英文)	2021—07	68.00	1384
复杂流的格子玻尔兹曼建模的工程应用(英文)	2021—08	68.00	1393
电偶极矩挑战(英文)	2021—08	108.00	1394
电动力学:问题与解法(英文)	2021—09	68.00	1395
自由电子激光的经典理论(英文)	2021—08	68.00	1397
曼哈顿计划——核武器物理学简介(英文)	2021—09	68.00	1401
粒子物理学(英文)	2021—09	68.00	1402
引力场中的量子信息(英文)	2021—09	128.00	1403
器件物理学的基本经典力学(英文)	2021—09	68.00	1404
等离子体物理及其空间应用导论.第1卷,基本原理和初步过程(英文)	2021—09	68.00	1405

书　　名	出版时间	定　价	编号
拓扑与超弦理论焦点问题(英文)	2021—07	58.00	1349
应用数学:理论、方法与实践(英文)	2021—07	78.00	1350
非线性特征值问题:牛顿型方法与非线性瑞利函数(英文)	2021—07	58.00	1351
广义膨胀和齐性:利用齐性构造齐次系统的李雅普诺夫函数和控制律(英文)	2021—06	48.00	1352
解析数论焦点问题(英文)	2021—07	58.00	1353
随机微分方程:动态系统方法(英文)	2021—07	58.00	1354
经典力学与微分几何(英文)	2021—07	58.00	1355
负定相交形式流形上的瞬子模空间几何(英文)	2021—07	68.00	1356

书　　名	出版时间	定　价	编号
广义卡塔兰轨道分析:广义卡塔兰轨道计算数字的方法(英文)	2021—07	48.00	1367
洛伦兹方法的变分:二维与三维洛伦兹方法(英文)	2021—08	38.00	1378
几何、分析和数论精编(英文)	2021—08	68.00	1380
从一个新角度看数论:通过遗传方法引入现实的概念(英文)	2021—07	58.00	1387
动力系统:短期课程(英文)	2021—08	68.00	1382
几何路径:理论与实践(英文)	2021—08	48.00	1385
论天体力学中某些问题的不可积性(英文)	2021—07	88.00	1396
广义斐波那契数列及其性质(英文)	2021—08	38.00	1386
对称函数和麦克唐纳多项式:余代数结构与 Kawanaka 恒等式(英文)	2021—09	38.00	1400

书　　名	出版时间	定　价	编号
杰弗里·英格拉姆·泰勒科学论文集:第1卷.固体力学(英文)	2021—05	78.00	1360
杰弗里·英格拉姆·泰勒科学论文集:第2卷.气象学、海洋学和湍流(英文)	2021—05	68.00	1361
杰弗里·英格拉姆·泰勒科学论文集:第3卷.空气动力学以及落弹数和爆炸的力学(英文)	2021—05	68.00	1362
杰弗里·英格拉姆·泰勒科学论文集:第4卷.有关流体力学(英文)	2021—05	58.00	1363

刘培杰数学工作室
已出版(即将出版)图书目录——原版影印

书名	出版时间	定价	编号
非局域泛函演化方程:积分与分数阶(英文)	2021—08	48.00	1390
理论工作者的高等微分几何:纤维丛、射流流形和拉格朗日理论(英文)	2021—08	68.00	1391
半线性退化椭圆微分方程:局部定理与整体定理(英文)	2021—07	48.00	1392
非交换几何、规范理论和重整化:一般简介与非交换量子场论的重整化(英文)	2021—09	78.00	1406
数论论文集:拉普拉斯变换和带有数论系数的幂级数(俄文)	2021—09	48.00	1407
挠理论专题:相对极大值,单射与扩充模(英文)	2021—09	88.00	1410
强正则图与欧几里得若尔当代数:非通常关系中的启示(英文)	2021—10	48.00	1411
拉格朗日几何和哈密顿几何:力学的应用(英文)	2021—10	48.00	1412
时滞微分方程与差分方程的振动理论:二阶与三阶(英文)	2021—10	98.00	1417
卷积结构与几何函数理论:用以研究特定几何函数理论方向的分数阶微积分算子与卷积结构(英文)	2021—10	48.00	1418
经典数学物理的历史发展(英文)	2021—10	78.00	1419
扩展线性丢番图问题(英文)	2021—10	38.00	1420
一类混沌动力系统的分歧分析与控制:分歧分析与控制(英文)	2021—11	38.00	1421
伽利略空间和伪伽利略空间中一些特殊曲线的几何性质(英文)	2022—01	68.00	1422
一阶偏微分方程:哈密尔顿—雅可比理论(英文)	2021—11	48.00	1424
各向异性黎曼多面体的反问题:分段光滑的各向异性黎曼多面体反边界谱问题:唯一性(英文)	2021—11	38.00	1425
项目反应理论手册.第一卷,模型(英文)	2021—11	138.00	1431
项目反应理论手册.第二卷,统计工具(英文)	2021—11	118.00	1432
项目反应理论手册.第三卷,应用(英文)	2021—11	138.00	1433
二次无理数:经典数论入门(英文)	2022—05	138.00	1434
数,形与对称性:数论,几何和群论导论(英文)	2022—05	128.00	1435
有限域手册(英文)	2021—11	178.00	1436
计算数论(英文)	2021—11	148.00	1437
拟群与其表示简介(英文)	2021—11	88.00	1438
数论与密码学导论:第二版(英文)	2022—01	148.00	1423

刘培杰数学工作室
已出版(即将出版)图书目录——原版影印

书名	出版时间	定价	编号
几何分析中的柯西变换与黎兹变换:解析调和容量和李普希兹调和容量、变化和振荡以及一致可求长性(英文)	2021—12	38.00	1465
近似不动点定理及其应用(英文)	2022—05	28.00	1466
局部域的相关内容解析:对局部域的扩展及其伽罗瓦群的研究(英文)	2022—01	38.00	1467
反问题的二进制恢复方法(英文)	2022—03	28.00	1468
对几何函数中某些类的各个方面的研究:复变量理论(英文)	2022—01	38.00	1469
覆盖、对应和非交换几何(英文)	2022—01	28.00	1470
最优控制理论中的随机线性调节器问题:随机最优线性调节器问题(英文)	2022—01	38.00	1473
正交分解法:涡流流体动力学应用的正交分解法(英文)	2022—01	38.00	1475
芬斯勒几何的某些问题(英文)	2022—03	38.00	1476
受限三体问题(英文)	2022—05	38.00	1477
利用马利亚万微积分进行 Greeks 的计算:连续过程、跳跃过程中的马利亚万微积分和金融领域中的 Greeks(英文)	2022—05	48.00	1478
经典分析和泛函分析的应用:分析学的应用(英文)	2022—03	38.00	1479
特殊芬斯勒空间的探究(英文)	2022—03	48.00	1480
某些图形的施泰纳距离的细谷多项式:细谷多项式与图的维纳指数(英文)	2022—05	38.00	1481
图论问题的遗传算法:在新鲜与模糊的环境中(英文)	2022—05	48.00	1482
多项式映射的渐近簇(英文)	2022—05	38.00	1483
一维系统中的混沌:符号动力学,映射序列,一致收敛和沙可夫斯基定理(英文)	2022—05	38.00	1509
多维边界层流动与传热分析:粘性流体流动的数学建模与分析(英文)	2022—05	38.00	1510
演绎理论物理学的原理:一种基于量子力学波函数的逐次置信估计的一般理论的提议(英文)	2022—05	38.00	1511
R^2 和 R^3 中的仿射弹性曲线:概念和方法(英文)	2022—08	38.00	1512
算术数列中除数函数的分布:基本内容、调查、方法、第二矩、新结果(英文)	2022—05	28.00	1513
抛物型狄拉克算子和薛定谔方程:不定常薛定谔方程的抛物型狄拉克算子及其应用(英文)	2022—07	28.00	1514
黎曼-希尔伯特问题与量子场论:可积重正化、戴森-施温格方程(英文)	2022—08	38.00	1515
代数结构和几何结构的形变理论(英文)	2022—08	48.00	1516
概率结构和模糊结构上的不动点:概率结构和直觉模糊度量空间的不动点定理(英文)	2022—08	38.00	1517

刘培杰数学工作室

已出版(即将出版)图书目录——原版影印

书　　名	出版时间	定　价	编号
反若尔当对:简单反若尔当对的自同构(英文)	2022-07	28.00	1533
对某些黎曼-芬斯勒空间变换的研究:芬斯勒几何中的某些变换(英文)	2022-07	38.00	1534
内诣零流形映射的尼尔森数的阿诺索夫关系(英文)	2023-01	38.00	1535
与广义积分变换有关的分数次演算:对分数次演算的研究(英文)	2023-01	48.00	1536
强子的芬斯勒几何和吕拉几何(宇宙学方面):强子结构的芬斯勒几何和吕拉几何(拓扑缺陷)(英文)	2022-08	38.00	1537
一种基于混沌的非线性最优化问题:作业调度问题(英文)	即将出版		1538
广义概率论发展前景:关于趣味数学与置信函数实际应用的一些原创观点(英文)	即将出版		1539

书　　名	出版时间	定　价	编号
纽结与物理学:第二版(英文)	2022-09	118.00	1547
正交多项式和q-级数的前沿(英文)	2022-09	98.00	1548
算子理论问题集(英文)	2022-09	108.00	1549
抽象代数:群、环与域的应用导论:第二版(英文)	即将出版		1550
菲尔兹奖得主演讲集:第三版(英文)	2023-01	138.00	1551
多元实函数教程(英文)	2022-09	118.00	1552
球面空间形式群的几何学:第二版(英文)	2022-09	98.00	1566

书　　名	出版时间	定　价	编号
对称群的表示论(英文)	2023-01	98.00	1585
纽结理论:第二版(英文)	2023-01	88.00	1586
拟群理论的基础与应用(英文)	2023-01	88.00	1587
组合学:第二版(英文)	2023-01	98.00	1588
加性组合学:研究问题手册(英文)	2023-01	68.00	1589
扭曲、平铺与镶嵌:几何折纸中的数学方法(英文)	2023-01	98.00	1590
离散与计算几何手册:第三版(英文)	2023-01	248.00	1591
离散与组合数学手册:第二版(英文)	2023-01	248.00	1592

书　　名	出版时间	定　价	编号
分析学教程.第1卷,一元实变量函数的微积分分析学介绍(英文)	2023-01	118.00	1595
分析学教程.第2卷,多元函数的微分和积分,向量微积分(英文)	2023-01	118.00	1596
分析学教程.第3卷,测度与积分理论,复变量的复值函数(英文)	2023-01	118.00	1597
分析学教程.第4卷,傅里叶分析,常微分方程,变分法(英文)	2023-01	118.00	1598

联系地址:哈尔滨市南岗区复华四道街10号　**哈尔滨工业大学出版社刘培杰数学工作室**

网　　址:http://lpj.hit.edu.cn/

邮　　编:150006

联系电话:0451-86281378　　13904613167

E-mail:lpj1378@163.com